Lecture Notes in Mathematics 1930

Editors:
J.-M. Morel, Cachan
F. Takens, Groningen
B. Teissier, Paris

José Miguel Urbano

The Method of Intrinsic Scaling

A Systematic Approach to Regularity for Degenerate and Singular PDEs

 Springer

Author

José Miguel Urbano
CMUC, Department of Mathematics
University of Coimbra
3001-454 Coimbra
Portugal
jmurb@mat.uc.pt

ISBN: 978-3-540-75931-7   e-ISBN: 978-3-540-75932-4
DOI: 10.1007/978-3-540-75932-4

Lecture Notes in Mathematics ISSN print edition: 0075-8434
                            ISSN electronic edition: 1617-9692

Library of Congress Control Number: 2008921371

Mathematics Subject Classification (2000): 35D10, 35K65

Cover design: WMXDesign GmbH

Printed on acid-free paper

9 8 7 6 5 4 3 2 1

springer.com

To Martim, Xavier and Catarina.

Preface

When I started giving talks on regularity theory for degenerate and singular parabolic equations, a fixed-point in the conversation during the coffee-break that usually followed the seminar was the apparent contrast between the beauty of the subject and its technical difficulty. I could not agree more on the beauty part but, most of the times, overwhelmingly failed to convince my audience that the technicalities were not all that hard to follow. As in many other instances, it was the fact that the results in the literature were eventually stated and proved in their most possible generality that made the whole subject seem inexpugnable.

So when I had the chance of preparing a short course on the method of intrinsic scaling, I decided to present the theory from scratch for the simplest model case of the degenerate p-Laplace equation and to leave aside technical refinements needed to deal with more general situations. The first part of the notes you are about to read is the result of that effort: an introductory and self-contained approach to intrinsic scaling, aiming at bringing to light what is really essential in this powerful tool in the analysis of degenerate and singular equations. As another striking feature of the method is its pervasiveness in terms of the applications, in the second part of the book, intrinsic scaling is applied to several models arising from flows in porous media, chemotaxis and phase transitions. The aim is to convince the reader of the strength of the method as a systematic approach to regularity for an important and relevant class of nonlinear partial differential equations.

The analysis of degenerate and singular parabolic equations is an extremely vast and active research topic and in this contribution there is, by no means, any intention to exhaust the theory. On the contrary, the focus is on a particular subject – the (Hölder) continuity of solutions – and a unifying set of ideas. We hope that the careful study of theses notes will enable the reader to master the essential features of the method of intrinsic scaling, which is instrumental in dealing with more elaborate aspects of the theory, like the boundedness of solutions, Harnack inequalities or systems of equations.

The first four chapters contain material that would fit well in an advanced graduate course on regularity theory for partial differential equations. Each chapter corresponds roughly, with the exception of the first one, to two 90 min. classes. Chapters 5–7 are independent from one another and each could be chosen to complement the course, according to individual preferences. I would probably suggest choosing chapter 5 for that purpose.

These lecture notes had its origin in a minicourse I delivered at the 2005 Summer Program of IMPA in Rio de Janeiro. Later that year, I taught a shorter version of the course at the University of Florence. I would like to thank Marcelo Viana and Vincenzo Vespri for their kind invitations and for the wonderful hospitality. I am also indebted to all the colleagues and students who took the course for their interest and input and, in particular, to my former PhD student Eurica Henriques.

Finally, I warmly thank Emmanuele DiBenedetto for his continuing support and advice.

Coimbra, November 2007   *José Miguel Urbano*

Contents

1

Introduction

Many relevant phenomena, not only in the natural sciences but also in engineering and economics, are modeled by (systems of) partial differential equations (PDEs) that exhibit some sort of degeneracy or singularity. Examples include the motion of multi-phase fluids in porous media, the melting of crushed ice (and phase transitions, in general), the behavior of composite materials or the pricing of assets in financial markets. Because of its significance in terms of the applications, but also due to the novel analytical techniques that it generates, the class of degenerate and singular parabolic equations is an important branch in the contemporary analysis of partial differential equations.

A central example, that will serve as a prototype along the text, is the parabolic p−Laplace equation

$$u_t - \operatorname{div} |\nabla u|^{p-2} \nabla u = 0, \qquad p > 1.$$

If $p > 2$, the equation is *degenerate* in the space part since its modulus of ellipticity $|\nabla u|^{p-2}$ vanishes at points where $|\nabla u| = 0$. If $1 < p < 2$, the modulus of ellipticity becomes unbounded at points where $|\nabla u| = 0$ and the equation is said to be *singular*.

In general, the nature and origin of the degeneracy or singularity may be quite different but it produces a common effect in the equation, namely the weakening of its structure and the prospect that some of the properties of its solutions be lost. From the point of view of regularity theory, the challenge is to understand to what extent this weakening of the structure, at the points where the PDE degenerates or becomes singular, compromises the regularizing effect that is typical of parabolic equations. Results on the continuity of weak solutions are paramount in this context since they assure that no singularities arise as a consequence of the degradation of the structure of the PDE.

The purpose of these lecture notes is to describe intrinsic scaling, a method for obtaining continuity results for the weak solutions of degenerate and singular parabolic equations, and to convince the reader of the strength of this approach to regularity, by giving evidence of its wide applicability in different situations. To understand what is at stake, let us start by placing the problem in its historical context. As in many other mathematical journeys, it all started with one of Hilbert's problems.

Hilbert's 19th Problem

In 1900, in a then obscure and now legendary session of the International Congress of Mathematicians in Paris, David Hilbert presented his list of 23 problems that would shape the mathematics of the newborn century. Two of those problems were related to the Calculus of Variations:

19th. *are the solutions of regular problems in the Calculus of Variations always necessarily analytic?*

20th. *do regular problems in the Calculus of Variations always possess a solution, (...) extending, if need be, the notion of solution?*

To be specific, suppose we want to minimize the functional

$$\mathcal{I}[w] = \int_{\Omega} f\left[\nabla w(x)\right] \, dx, \tag{1.1}$$

over all $w : \Omega \to \mathbb{R}$ such that $w = g$ on $\partial\Omega$. Here, $\Omega \subset \mathbb{R}^d$ and $f : \mathbb{R}^d \to \mathbb{R}$ is a given real function (possibly nonlinear), the so-called Lagrangian. The minimization problem is said to be regular if the Lagrangian $f(\zeta)$ is regular and convex. The problem of minimizing the functional (1.1) is associated with its Euler-Lagrange equation

$$\left(f_{\zeta_i}(\nabla u)\right)_{x_i} = 0 \quad \text{in} \quad \Omega,$$

solved by any minimizer. If one takes the Euler-Lagrange equation for a minimizer w^* and differentiate it with respect to x_k, the result is that, for any $k \in \{1, 2, \ldots, d\}$, the partial derivative $(w^*)_{x_k} := v_k$ solves the linear second-order partial differential equation

$$\left(a_{ij}(x) \, u_{x_j}\right)_{x_i} = 0 \quad \text{in} \quad \Omega, \tag{1.2}$$

with coefficients $a_{ij}(x) := f_{\zeta_i \zeta_j}(\nabla w^*(x))$. The equation is (uniformly) elliptic since f is (strictly) convex. The regularity problem for minimizers of convex functionals is thus closely related to the problem of the regularity of the solutions of elliptic equations.

In 1904, Bernstein proved that a C^3 solution of equation (1.2) is necessarily analytic, what was seen at the time as a solution to the problem. The subsequent progress focused on weakening the departing regularity required to derive the analyticity. All the results obtained (by E. Hopf, Morrey and Stampacchia, among others) were based, one way or the other, in perturbation arguments and the comparison of solutions with harmonic functions. As such, they all required some sort of regularity of the coefficients a_{ij} (at least continuity). Without the regularity of the coefficients – assuming, for example, they were merely bounded – it was not possible to establish the regularity of the solutions of equation (1.2), let alone that of the minimizers of (1.1). The available theory was based on Schauder estimates that, to simplify, placed the solutions of (1.2) in $C^{k+1,\alpha}$ if the coefficients a_{ij} belonged to $C^{k,\alpha}$. This supported the following iterative reasoning (known as *bootstrap* argument)

$$w^* \in C^{1,\alpha} \Rightarrow a_{ij} := f_{\zeta_i \zeta_j}(\nabla w^*) \in C^{0,\alpha} \Rightarrow v_k \in C^{1,\alpha}$$
$$\Downarrow$$
$$v_k \in C^{2,\alpha} \Leftarrow a_{ij} := f_{\zeta_i \zeta_j}(\nabla w^*) \in C^{1,\alpha} \Leftarrow w^* \in C^{2,\alpha}$$
$$\Downarrow$$
$$w^* \in C^{3,\alpha} \Rightarrow \qquad \cdots$$

so ultimately $w^* \in C^\infty$. The analyticity then followed.

Meanwhile, the existence theory for the Calculus of Variations started developing through the use of direct methods. Imposing, in addition to the convexity, a coercivity assumption on the Lagrangian (of the type $|f(\zeta)| \geq \delta |\zeta|^2 - \beta$, $\zeta \in \mathbb{R}^d$), it was possible to guarantee the existence of a minimizer, provided that the notion of solution was adequately extended – which was entirely in the spirit of the formulation of Hilbert's 20th problem. The appropriate class of admissible functions, for the corresponding minimization problem, turned out to be the Sobolev space H^1, *i.e.*, the space of L^2 functions whose first derivatives (in the sense of distributions) still belong to L^2.

There was thus a gap between the existence and regularity classes. Extending the notion of solution, it was possible to obtain the existence of a minimizer in H^1. From this, the measurability and, with natural hypothesis on the Lagrangian ($|D^2 f(\zeta)| \leq C$, $\zeta \in \mathbb{R}^d$), the boundedness of the a_{ij} followed. The question, in the early 1950's, boiled down to showing that the solutions of (1.2), with coefficients merely measurable and bounded, were Hölder continuous:

$$w^* \in H^1 \Rightarrow a_{ij} := f_{\zeta_i \zeta_j}(\nabla w^*) \in L^\infty \overset{?}{\Rightarrow} v_k \in C^{0,\alpha} \Rightarrow w^* \in C^{1,\alpha}$$

It was, arguably, the most important problem in the analysis of partial differential equations at that time. The problem had been solved by Morrey, in 1938 [41], but only for $d = 2$, and the techniques used were typically bidimensional, involving complex analysis and quasi-conformal mappings. In the geral case, only Stampacchia's estimates were known, giving $u \in H^2$; the problem remained widely open.

Enters De Giorgi

In 1956, Ennio De Giorgi, a young mathematician of Pisa, then twenty-nine, submits to the *Rendiconti dell'Accademia Nazionale dei Lincei* a four pages

note, titled *Sull'analiticità degli estremali degli integrali multipli*, announcing
the proof (later published in [10]) of the Hölder continuity of the functions
that satisfy certain integral conditions – later known as elements of De Giorgi's
class. As the solutions of (1.2), with measurable and bounded coefficients, belong to that class, a positive answer to Hilbert's 19th problem, in the scalar
case and any dimension, followed as an immediate consequence. In the vec-
torial case, the answer is negative and it was again De Giorgi, around ten
years later, who constructed a counter-example of a discontinuous solution to
a regular (in the sense of Hilbert) variational system.

De Giorgi uses a completely original approach to obtain the *a priori* that
lead to the solution of the problem. The following is Bombieri's testimony [5]
(translated freely from the Italian):

*Once I asked De Giorgi how he got to the idea that led him to solve this problem.
He replied as if it was all an indirect consequence of another problem, much more
difficult, that he was studying at that moment, namely the isoperimeteric problem in
several dimensions, and started explaining the connections between the two problems.
I then realized that De Giorgi looked at these functions of several variables literally
as geometric objects in space. In his explanation, he kept moving his hands as if he
was touching an invisible surface, and showing how to perform his operations and
transformations, cutting and pasting invisible masses from one side to the other,
leveling and filling the peaks and valleys of theses surfaces. In the specific case, it
consisted of taking the level curves of the surface that solved the problem and applying
his isoperimetric property. To me, it was an usual way of doing analysis, a field that
often requires the use of rather fine estimates, that the normal mathematician grasps
more easily through the formulas than through the geometry.*

The work of De Giorgi concerns uniformly elliptic linear equations but
the linearity does not play a role in the proofs. This allowed Ladyzhenskaya
and Uralt'seva [38] to extend, in the mid 1960's, the results on the Hölder
continuity of weak solutions to quasi-linear equations of the form

$$\text{div } \mathbf{a}(x, u, \nabla u) = 0 \quad \text{in } \Omega, \tag{1.3}$$

with structure assumptions of the type

$$\begin{cases} \mathbf{a}(x, u, \nabla u) \cdot \nabla u \geq C_0 \, |\nabla u|^p - C \\[2mm] |\mathbf{a}(x, u, \nabla u)| \leq C \left(|\nabla u|^{p-1} + 1\right), \end{cases} \tag{1.4}$$

with $p > 1$, and constants $C_0 > 0$ and $C \geq 0$. The generalization is twofold:
the principal part $\mathbf{a}(x, u, \nabla u)$ is permitted to have a nonlinear dependence
in ∇u, and a *non-linear growth* with respect to $|\nabla u|$. The latter is of partic-
ular interest since equation (1.3) might be either degenerate or singular, as
illustrated by the archetypal $p-$Laplace equation

$$\text{div } |\nabla u|^{p-2}\nabla u = 0.$$

From Elliptic to Parabolic

Let us now consider the parabolic analogues of equations (1.2) and (1.3), that is, with $\Omega_T = \Omega \times (0, T]$, $0 < T < \infty$,

$$u_t - \left(a_{ij}(x, t)\, u_{x_j}\right)_{x_i} = 0 \quad \text{in} \quad \Omega_T, \tag{1.5}$$

with bounded and measurable coefficients a_{ij} satisfying an ellipticity condition, and

$$u_t - \operatorname{div} \mathbf{a}(x, t, u, \nabla u) = 0 \quad \text{in} \quad \Omega_T, \tag{1.6}$$

with $\mathbf{a}$ satisfying structure assumptions analogous to (1.4). Moser [43] proved that weak solutions of (1.5) are locally Hölder continuous in Ω_T. Since the linearity is immaterial to the proof, one might expect, as in the elliptic case, an extension of these results to quasi-linear equations of the type (1.6), where the structure condition is as in (1.4). Surprisingly though, the methods of De Giorgi and Moser could not be extended. Ladyzhenskaya *et als.* [37] proved that solutions of (1.6) are Hölder continuous, provided the principal part has exactly a linear growth with respect to $|\nabla u|$, *i.e.*, if $p = 2$ in the structure assumptions corresponding to (1.4). Analogous results were established by Kruzkov [33, 34] and by Nash [44] using entirely different methods. Thus it appears that unlike the elliptic case, the degeneracy or singularity of the principal part plays a peculiar role, and for example, for the parabolic p–Laplace equation

$$u_t - \operatorname{div} |\nabla u|^{p-2} \nabla u = 0$$

one could not establish whether a solution is locally Hölder continuous.

Intrinsic Scaling: A New Approach to Regularity

The issue remained open until the mid 1980's when DiBenedetto [12] showed that the solutions of general quasilinear equations of the type of (1.6) are locally Hölder continuous for $p > 2$. In the early 1990's, the theory was extended [9] to include also the case $1 < p < 2$. Surprisingly, the same techniques could be suitably modified to establish the local Hölder continuity of any local solution of quasilinear porous medium-type equations. These modified methods, in turn, were crucial in proving that weak solutions of the p–Laplace equation are of class $C^{1,\alpha}_{\mathrm{loc}}$.

  These results follow, one way or another, from the single unifying idea of *intrinsic scaling*: the diffusion processes in the equations evolve in a time scale determined instant by instant by the solution itself, so that, loosely speaking, they can be regarded as the heat equation in their own intrinsic time-configuration. A precise description of this fact, as well as its effectiveness, is linked to its technical implementation, which we will develop in this set of lecture notes.

The continuity of a solution at a point follows from measuring its oscillation in a sequence of nested and shrinking cylinders, with vertex at that point, and showing that the oscillation converges to zero as the cylinders shrink to the point. When it is possible to describe quantitatively how the oscillation converges, a modulus of continuity is derived. The idea behind the method of intrinsic scaling is to perform this iterative process in cylinders that reflect the structure of the equation. By this, we mean cylinders whose dimensions, with respect to the standard parabolic cylinders, are redefined in terms of scaling factors that take into account the nature of the degeneracy or singularity, and depend on the oscillation of the solution itself (thus, the term intrinsic). The cylinders should trivially reduce to the standard parabolic cylinders, reflecting the natural homogeneity between the space and time variables for the heat equation, in the particular case of a uniformly parabolic equation. The punch line of the theory is that *the equation behaves, in its own geometry, like the heat equation.*

The building blocks of the method of intrinsic scaling are *a priori* estimates for the weak solutions of the equation. Actually, there is more to it than that. Once these estimates are obtained, we can forget the equation and the problem becomes, purely, a problem in analysis: showing that functions that satisfy certain integral inequalities belong to a certain regularity class (*e.g.*, are locally Hölder continuous). It does not really matter if these functions are solutions of an equation or extremals of a functional in the Calculus of Variations or neither of those; what counts is that they satisfy the integral estimates. These integral inequalities on level sets *measure* the behaviour of the function near its infimum and its supremum in the interior of the cylinder. In the case of solutions of degenerate or singular equations, these estimates are not homogeneous since they involve integral norms corresponding to different powers. This lack of homogeneity precludes the use of certain functional inclusions because the appropriate norms are not disclosed in the analysis. Through intrinsic scaling, we are able to recover the homogeneity in the estimates, once we rewrite them over the intrinsically rescaled cylinders. The difficulty in the analysis is thus absorbed by the geometry.

Emerging Trends

In a series of papers [17, 18, 19], DiBenedetto, Gianazza and Vespri established recently an intrinsic Harnack inequality for nonnegative solutions of degenerate and singular parabolic PDEs with the full quasi-linear structure, a problem that remained open for about 40 years. The result, that provides an alternative and independent proof of the Hölder continuity of solutions, was known only for the particular case of the parabolic p-Laplacian, and the available proof was clearly of a restricted applicability since it used, in an essential way, the maximum principle and comparison with p-heat potentials. In a way, this new result parallels the celebrated work [43] of Moser in the 1960's, when the Harnack inequality for general nondegenerate ($p = 2$) quasilinear parabolic

PDEs was obtained. The true advance in Moser's proof was also in bypassing the use of potentials that were essential in the approach of Hadamard [25] and Pini [45] concerning the heat equation. There is even a further merit in the contribution of DiBenedetto *et als*: unlike Moser's proof, no use is made of some fine analytical properties of BMO spaces (the use of John-Nirenberg lemma was rather complicated in the original paper of Moser, who provided a simpler proof some time later). The arguments are measure-theoretical in nature and, as the authors remark, *hold the promise of a wider applicability*. See also [36] for a different perspective in the degenerate case.

For more on contemporary issues related to regularity estimates for degenerate and singular elliptic and parabolic equations, we address the interested reader to the recent special issue [20], where several contributions are collected.

Outline of the Lecture Notes

We have decided to present the theory for the model case of the degenerate p–Laplace equation to bring to light what is really essential in the method, leaving aside technical refinements needed to deal with more general equations.

In chapter 2, the precise definition of weak solution is introduced for the model problem, and we derive local energy and logarithmic estimates, the building blocks of the theory. Chapter 3 deals with the construction of the appropriate geometric setting, bringing about in full detail the idea of intrinsic scaling; it also highlights the precise role of the logarithmic estimates. The 4th chapter culminates with the proof of the Hölder continuity, after the analysis of an alternative aimed at reducing the oscillation of the solution. The last section contains a series of generalizations, mainly to equations with the full quasilinear structure and of porous medium type.

This first part is essentially an edited version of the second section of [21]; some proofs are presented in more detail but the overlapping of the material is substantial. The division into sections and the order in which the issues are treated is rather different though, and it is intended to further simplify the reading.

The second part is devoted to a series of three applications of the theory to relevant models arising from flows in porous media, chemotaxis and phase transitions. In chapter 5, the flow of two immiscible fluids through a porous medium is studied and intrinsic scaling is used to obtain the Hölder continuity of the saturation, that satisfies a PDE with a two-sided degeneracy. The same type of structure arises in a model of chemotaxis with volume–filling effect and the extension to this case, which basically consists in dealing appropriately with an extra lower order term, is also included. In chapter 6, we obtain the continuity of the weak solutions of the porous medium equation with a variable exponent, generalizing the classical result to an increasingly popular context. Finally, in chapter 7, the Stefan problem for the singular p-Laplacian is considered and intrinsic scaling is again used to derive the continuity of the temperature, showing that no jumps occur across the free boundary.

These three examples were originally treated in [53, 4], [28] and [27], respectively. The contents of the chapters reflect this fact, with simplifications and extra remarks whenever appropriate. The novelty here lies in the unified approach, intended to convince the reader of the strength of the method of intrinsic scaling as a systematic approach to regularity for degenerate and singular PDEs.

Part I

The Method of Intrinsic Scaling

2

Weak Solutions and *a Priori* Estimates

We will concentrate on the parabolic p–Laplace equation

$$u_t - \operatorname{div} |\nabla u|^{p-2} \nabla u = 0 , \qquad p > 1, \qquad (2.1)$$

a quasilinear second-order partial differential equation, with principal part in divergence form. If $p > 2$, the equation is degenerate in the space part, due to the vanishing of its modulus of ellipticity $|\nabla u|^{p-2}$ at points where $|\nabla u| = 0$. The singular case corresponds to $1 < p < 2$: the modulus of ellipticity becomes unbounded at points where $|\nabla u| = 0$.

In this chapter we place no restriction on the values of $p > 1$. The theory is markedly different in the degenerate and singular cases and we will later restrict our attention to $p > 2$. The results extend to a variety of equations and, in particular, to equations with general principal parts satisfying appropriate structure assumptions and with lower order terms. We have chosen to present the results and the proofs for the particular model case (2.1) to bring to light what we feel are the essential features of the theory. Remarks on generalizations, which in some way or another correspond to more or less sophisticated technical improvements, are left to a later section.

2.1 Definition of Weak Solution

Let Ω be a bounded domain in $\mathbb{R}^d$, with smooth boundary $\partial\Omega$. Let

$$\Omega_T = \Omega \times (0, T] , \quad T > 0,$$

be the space-time domain, with lateral boundary $\Sigma = \partial\Omega \times (0, T)$ and parabolic boundary

$$\partial_p \Omega_T = \Sigma \cup (\Omega \times \{0\}) .$$

We start with the precise definition of local weak solution for (2.1).

Definition 2.1. *A local weak solution of* (2.1) *is a measurable function*

$$u \in C_{\text{loc}}\left(0, T; L^2_{\text{loc}}(\Omega)\right) \cap L^p_{\text{loc}}\left(0, T; W^{1,p}_{\text{loc}}(\Omega)\right)$$

such that, for every compact $K \subset \Omega$ and for every subinterval $[t_1, t_2]$ of $(0, T]$,

$$\int_K u\varphi \, dx \Big|_{t_1}^{t_2} + \int_{t_1}^{t_2} \int_K \left\{ -u\varphi_t + |\nabla u|^{p-2}\nabla u \cdot \nabla \varphi \right\} dx \, dt = 0, \qquad (2.2)$$

for all $\varphi \in H^1_{\text{loc}}\left(0, T; L^2(K)\right) \cap L^p_{\text{loc}}\left(0, T; W^{1,p}_0(K)\right)$.

It would be technically convenient to have at hand a formulation of weak solution involving the time derivative u_t. Unfortunately, solutions of (2.1), whenever they exist, possess a modest degree of time-regularity and, in general, u_t has a meaning only in the sense of distributions. To overcome this limitation, we introduce the Steklov average of a function $v \in L^1(\Omega_T)$, defined, for $0 < h < T$, by

$$v_h := \begin{cases} \dfrac{1}{h} \displaystyle\int_t^{t+h} v(\cdot, \tau) \, d\tau & \text{if } t \in (0, T - h] \\[2ex] 0 & \text{if } t \in (T - h, T]. \end{cases} \qquad (2.3)$$

The proof of the following lemma follows from the general theory of L^p spaces.

Lemma 2.2. *If $v \in L^{q,r}(\Omega_T)$ then, as $h \to 0$, the Steklov average v_h converges to v in $L^{q,r}(\Omega_{T-\epsilon})$, for every $\epsilon \in (0, T)$. If $v \in C\left(0, T; L^q(\Omega)\right)$ then, as $h \to 0$, the Steklov average $v_h(\cdot, t)$ converges to $v(\cdot, t)$ in $L^q(\Omega)$, for every $t \in (0, T-\epsilon)$ and every $\epsilon \in (0, T)$.*

It is a simple exercise to show that the definition of local weak solution previously introduced is equivalent to the following one.

Definition 2.3. *A local weak solution of* (2.1) *is a measurable function*

$$u \in C_{\text{loc}}\left(0, T; L^2_{\text{loc}}(\Omega)\right) \cap L^p_{\text{loc}}\left(0, T; W^{1,p}_{\text{loc}}(\Omega)\right)$$

such that, for every compact $K \subset \Omega$ and for every $0 < t < T - h$,

$$\int_{K \times \{t\}} \left\{ (u_h)_t \, \varphi + \left(|\nabla u|^{p-2}\nabla u\right)_h \cdot \nabla \varphi \right\} dx = 0, \qquad (2.4)$$

for all $\varphi \in W^{1,p}_0(K)$.

We will show that locally bounded solutions of (2.1) are locally Hölder continuous within their domain of definition. No specific boundary or initial values need to be prescribed for u. A theory of boundedness of weak solutions of (2.1) is quite different from the linear theory (*cf.* [14]): weak solutions are locally bounded only if $d(p-2) + p > 0$. It can be shown by counterexample that this condition is sharp. Although the arguments below are of local nature, to simplify the presentation we assume that u is a.e. defined and bounded in Ω_T and set

$$M := \|u\|_{L^\infty(\Omega_T)}.$$

2.2 Local Energy Estimates: The Building Blocks of the Theory

The building blocks of the method of intrinsic scaling are *a priori* estimates for weak solutions. Once these estimates are obtained, we can forget the equation and the problem becomes, purely, a problem in analysis: showing that functions that satisfy certain integral inequalities belong to a certain regularity class (*e.g.*, are locally Hölder continuous). These estimates are integral inequalities on level sets that *measure* the behaviour of the function near its infimum and its supremum in the interior of an appropriate cylinder.

Given a point $x_0 \in \mathbb{R}^d$, denote by $K_\rho(x_0)$ the d-dimensional cube with centre at x_0 and wedge 2ρ:

$$K_\rho(x_0) := \left\{ x \in \mathbb{R}^d : \max_{1 \le i \le d} |x_i - x_{0i}| < \rho \right\}$$

and put $K_\rho := K_\rho(0)$; given a point $(x_0, t_0) \in \mathbb{R}^{d+1}$, the cylinder of radius ρ and height $\tau > 0$ with vertex at (x_0, t_0) is

$$(x_0, t_0) + Q(\tau, \rho) := K_\rho(x_0) \times (t_0 - \tau, t_0).$$

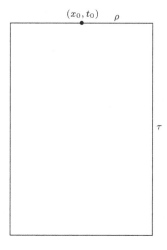

We write $Q(\tau, \rho)$ to denote $(0,0) + Q(\tau, \rho)$. We use the usual notations for the positive and negative parts of a function:

$$v_+ = \max(v, 0) \qquad \text{and} \qquad v_- = (-v)_+.$$

We now deduce the energy estimates. Without loss of generality, we restrict to cylinders with vertex at the origin $(0,0)$, the changes being obvious for cylinders with vertex at a generic (x_0, t_0). Consider a cylinder $Q(\tau, \rho) \subset \Omega_T$ and let $0 \leq \zeta \leq 1$ be a piecewise smooth cutoff function in $Q(\tau, \rho)$ such that

$$|\nabla \zeta| < \infty \qquad \text{and} \qquad \zeta(x, t) = 0, \quad x \notin K_\rho. \tag{2.5}$$

Proposition 2.4. *Let u be a local weak solution of (2.1) and $k \in \mathbb{R}$. There exists a constant $C \equiv C(p) > 0$ such that, for every cylinder $Q(\tau, \rho) \subset \Omega_T$,*

$$\sup_{-\tau < t < 0} \int_{K_\rho \times \{t\}} (u - k)_\pm^2 \, \zeta^p \, dx + \int_{-\tau}^0 \int_{K_\rho} |\nabla (u - k)_\pm \zeta|^p \, dx \, dt$$

$$\leq \int_{K_\rho \times \{-\tau\}} (u - k)_\pm^2 \, \zeta^p \, dx + C \int_{-\tau}^0 \int_{K_\rho} (u - k)_\pm^p \, |\nabla \zeta|^p \, dx \, dt$$

$$+ p \int_{-\tau}^0 \int_{K_\rho} (u - k)_\pm^2 \, \zeta^{p-1} \, \zeta_t \, dx \, dt. \tag{2.6}$$

Proof. Let $\varphi = \pm(u_h - k)_\pm \zeta^p$ in (2.4) and integrate in time over $(-\tau, t)$ for $t \in (-\tau, 0)$. The first term gives

$$\int_{-\tau}^t \int_{K_\rho} (u_h)_t \, \varphi \, dx \, d\theta = \frac{1}{2} \int_{-\tau}^t \int_{K_\rho} \left[(u_h - k)_\pm^2 \right]_t \zeta^p \, dx \, d\theta$$

$$\longrightarrow \frac{1}{2} \int_{K_\rho \times \{t\}} (u - k)_\pm^2 \, \zeta^p \, dx - \frac{1}{2} \int_{K_\rho \times \{-\tau\}} (u - k)_\pm^2 \, \zeta^p \, dx$$

$$- \frac{p}{2} \int_{-\tau}^t \int_{K_\rho} (u - k)_\pm^2 \, \zeta^{p-1} \, \zeta_t \, dx \, d\theta,$$

after integrating by parts and passing to the limit in $h \to 0$ (using Lemma 2.2). Concerning the other term, letting first $h \to 0$, we obtain

$$\int_{-\tau}^t \int_{K_\rho} \left[|\nabla u|^{p-2} \nabla u \right]_h \cdot \nabla \varphi \, dx \, d\theta$$

$$\longrightarrow \int_{-\tau}^t \int_{K_\rho} |\nabla u|^{p-2} \nabla u \cdot \left[\pm \nabla (u - k)_\pm \zeta^p \pm p(u - k)_\pm \zeta^{p-1} \nabla \zeta \right] dx \, d\theta$$

$$\geq \int_{-\tau}^t \int_{K_\rho} |\nabla (u - k)_\pm|^p \, \zeta^p \, dx \, d\theta$$

$$-p \int_{-\tau}^{t} \int_{K_\rho} |\nabla(u-k)_\pm|^{p-1} (u-k)_\pm \, \zeta^{p-1} \, |\nabla\zeta| \, dx \, d\theta$$

$$\geq \frac{1}{2} \int_{-\tau}^{t} \int_{K_\rho} |\nabla(u-k)_\pm \, \zeta|^p \, dx \, d\theta$$

$$-C(p) \int_{-\tau}^{t} \int_{K_\rho} (u-k)_\pm^p \, |\nabla\zeta|^p \, dx \, d\theta,$$

using the inequality of Young

$$ab \leq \frac{\varepsilon^p}{p} a^p + \frac{1}{p'\varepsilon^{p'}} b^{p'},$$

with the choices

$$a = (u-k)_\pm |\nabla\zeta| \, , \quad b = |\nabla(u-k)_\pm \, \zeta|^{p-1} \, , \quad \text{and} \quad \varepsilon = [2(p-1)]^{\frac{1}{p'}}.$$

Since $t \in (-\tau, 0)$ is arbitrary, we can combine both estimates to obtain (2.6).

$\square$

Remark 2.5. In (2.6), there is an intentional ambiguity in the way we wrote $|\nabla(u-k)_\pm\zeta|^p$. The gradient can either affect only $(u-k)_\pm$ (as follows directly from the estimates in the proof) or the product $(u-k)_\pm\zeta$ (as the extra term can clearly be absorbed into the right hand side of the estimate).

2.3 Local Logarithmic Estimates

We now introduce a logarithmic function for which we obtain further local estimates. These are the subsidiary building blocks of the theory but nevertheless play a crucial role in the proof, allowing for the expansion in time to a full cylinder $Q(\tau, \rho)$ of certain results obtained for sub-cylinders of $Q(\tau, \rho)$.

Given constants a, b, c, with $0 < c < a$, define the nonnegative function

$$\psi_{\{a,b,c\}}^\pm (s) := \left(\ln \left\{ \frac{a}{(a+c) - (s-b)_\pm} \right\} \right)_+$$

$$= \begin{cases} \ln \left\{ \frac{a}{(a+c)\pm(b-s)} \right\} & \text{if } b \pm c \lessgtr s \lessgtr b \pm (a+c) \\ 0 & \text{if } s \lesseqgtr b \pm c \end{cases}$$

whose first derivative is

$$\left(\psi_{\{a,b,c\}}^\pm \right)' (s) = \begin{cases} \dfrac{1}{(b-s) \pm (a+c)} & \text{if } b \pm c \lessgtr s \lessgtr b \pm (a+c) \\ 0 & \text{if } s \lessgtr b \pm c \end{cases} \gtreqless 0,$$

and second derivative, off $s = b \pm c$, is

$$\left(\psi^{\pm}_{\{a,b,c\}}\right)'' = \left\{\left(\psi^{\pm}_{\{a,b,c\}}\right)'\right\}^2 \geq 0.$$

Now, given a bounded function u in a cylinder $(x_0, t_0) + Q(\tau, \rho)$ and a number k, define the constant

$$H^{\pm}_{u,k} := \operatorname*{ess\ sup}_{(x_0,t_0)+Q(\tau,\rho)} |(u-k)_{\pm}|.$$

The following function was introduced in [11] and since then has been used as a recurrent tool in the proof of results concerning the local behaviour of solutions of degenerate and singular equations:

$$\Psi^{\pm}\left(H^{\pm}_{u,k}, (u-k)_{\pm}, c\right) \equiv \psi^{\pm}_{\{H^{\pm}_{u,k}, k, c\}}(u), \qquad 0 < c < H^{\pm}_{u,k}. \qquad (2.7)$$

From now on, when referring to this function we will write it as $\psi^{\pm}(u)$, omitting the subscripts, whose meaning will be clear from the context.

Let $x \mapsto \zeta(x)$ be a time-independent cutoff function in $K_\rho(x_0)$ satisfying (2.5). The logarithmic estimates in cylinders $Q(\tau, \rho)$ with vertex at the origin read as follows.

Proposition 2.6. *Let u be a local weak solution of (2.1) and $k \in \mathbb{R}$. There exists a constant $C \equiv C(p) > 0$ such that, for every cylinder $Q(\tau, \rho) \subset \Omega_T$,*

$$\sup_{-\tau < t < 0} \int_{K_\rho \times \{t\}} \left[\psi^{\pm}(u)\right]^2 \zeta^p \, dx \leq \int_{K_\rho \times \{-\tau\}} \left[\psi^{\pm}(u)\right]^2 \zeta^p \, dx$$

$$+ C \int_{-\tau}^{0} \int_{K_\rho} \psi^{\pm}(u) \left|(\psi^{\pm})'(u)\right|^{2-p} |\nabla \zeta|^p \, dx \, dt. \qquad (2.8)$$

Proof. Take $\varphi = 2\,\psi^{\pm}(u_h) \left[(\psi^{\pm})'(u_h)\right] \zeta^p$ as a testing function in (2.4) and integrate in time over $(-\tau, t)$ for $t \in (-\tau, 0)$. Since $\zeta_t \equiv 0$,

$$\int_{-\tau}^{t} \int_{K_\rho} (u_h)_t \left\{2\,\psi^{\pm}(u_h) \left[(\psi^{\pm})'(u_h)\right] \zeta^p\right\} dx \, d\theta$$

$$= \int_{-\tau}^{t} \int_{K_\rho} \left(\left[\psi^{\pm}(u_h)\right]^2\right)_t \zeta^p \, dx \, d\theta$$

$$= \int_{K_\rho \times \{t\}} \left[\psi^{\pm}(u_h)\right]^2 \zeta^p \, dx - \int_{K_\rho \times \{-\tau\}} \left[\psi^{\pm}(u_h)\right]^2 \zeta^p \, dx.$$

From this, letting $h \to 0$,

$$\int_{-\tau}^{t} \int_{K_\rho} (u_h)_t \left\{ 2 \, \psi^\pm(u_h) \left[(\psi^\pm)'(u_h) \right] \zeta^p \right\} dx \, d\theta$$

$$\longrightarrow \int_{K_\rho \times \{t\}} \left[\psi^\pm(u) \right]^2 \zeta^p \, dx - \int_{K_\rho \times \{-\tau\}} \left[\psi^\pm(u) \right]^2 \zeta^p \, dx.$$

As for the remaining term, we first let $h \to 0$, to obtain

$$\int_{-\tau}^{t} \int_{K_\rho} |\nabla u|^{p-2} \nabla u \cdot \nabla \left\{ 2 \, \psi^\pm(u) \left[(\psi^\pm)'(u) \right] \zeta^p \right\} dx \, d\theta$$

$$= \int_{-\tau}^{t} \int_{K_\rho} |\nabla u|^p \left\{ 2 \left(1 + \psi^\pm(u) \right) \left[(\psi^\pm)'(u) \right]^2 \zeta^p \right\} dx \, d\theta$$

$$+ p \int_{-\tau}^{t} \int_{K_\rho} |\nabla u|^{p-2} \nabla u \cdot \nabla \zeta \left\{ 2 \, \psi^\pm(u) \left[(\psi^\pm)'(u) \right] \zeta^{p-1} \right\} dx \, d\theta$$

$$\geq \int_{-\tau}^{t} \int_{K_\rho} |\nabla u|^p \left\{ 2 \left(1 + \psi^\pm(u) - \psi^\pm(u) \right) \left[(\psi^\pm)'(u) \right]^2 \zeta^p \right\} dx \, d\theta$$

$$- 2(p-1)^{p-1} \int_{-\tau}^{t} \int_{K_\rho} \psi^\pm(u) \left| (\psi^\pm)'(u) \right|^{2-p} |\nabla \zeta|^p \, dx \, d\theta$$

$$\geq - C \int_{-\tau}^{t} \int_{K_\rho} \psi^\pm(u) \left| (\psi^\pm)'(u) \right|^{2-p} |\nabla \zeta|^p \, dx \, d\theta.$$

Since $t \in (-\tau, 0)$ is arbitrary, we can combine both estimates to obtain (2.8).

$\square$

2.4 Some Technical Tools

We gather in this section a few technical facts that, although marginal to the theory, are essential in establishing its main results.

1. A Lemma of De Giorgi

Given a continuous function $v : \Omega \to \mathbb{R}$ and two real numbers $k_1 < k_2$, we define

$$\begin{aligned}
[v > k_2] &:= \{ x \in \Omega \; : \; v(x) > k_2 \}, \\
[v < k_1] &:= \{ x \in \Omega \; : \; v(x) < k_1 \}, \\
[k_1 < v < k_2] &:= \{ x \in \Omega \; : \; k_1 < v(x) < k_2 \}.
\end{aligned} \tag{2.9}$$

Lemma 2.7 (De Giorgi, [10]). *Let* $v \in W^{1,1}\left(B_\rho(x_0)\right) \cap C\left(B_\rho(x_0)\right)$, *with* $\rho > 0$ *and* $x_0 \in \mathbb{R}^d$, *and let* $k_1 < k_2 \in \mathbb{R}$. *There exists a constant* C, *depending only on* d *(and thus independent of* ρ, x_0, v, k_1 *and* k_2), *such that*

$$(k_2 - k_1)\, |[v > k_2]| \leq C\, \frac{\rho^{d+1}}{|[v < k_1]|} \int_{[k_1 < v < k_2]} |\nabla v|\, dx.$$

Remark 2.8. The conclusion of the lemma remains valid, provided Ω is convex, for functions $v \in W^{1,1}(\Omega) \cap C(\Omega)$. We will use it in the case Ω is a cube. In fact, the continuity is not essential for the result to hold. For a function merely in $W^{1,1}(\Omega)$, we define the sets in (2.9) through any representative in the equivalence class. It can be shown that the conclusion of the lemma is independent of that choice.

2. Geometric Convergence of Sequences

The following lemmas concern the geometric convergence of sequences and are instrumental in the iterative schemes that will be derived along the proofs.

Lemma 2.9. *Let* (X_n), $n = 0, 1, 2, \ldots$, *be a sequence of positive real numbers satisfying the recurrence relation*

$$X_{n+1} \leq C\, b^n\, X_n^{1+\alpha}$$

where $C, b > 1$ *and* $\alpha > 0$ *are given. If*

$$X_0 \leq C^{-1/\alpha}\, b^{-1/\alpha^2}$$

then $X_n \to 0$ *as* $n \to \infty$.

Lemma 2.10. *Let* (X_n) *and* (Z_n), $n = 0, 1, 2, \ldots$, *be sequences of positive real numbers satisfying the recurrence relations*

$$\begin{cases} X_{n+1} \leq C\, b^n \left(X_n^{1+\alpha} + X_n^\alpha Z_n^{1+\kappa}\right) \\[2mm] Z_{n+1} \leq C\, b^n \left(X_n + Z_n^{1+\kappa}\right) \end{cases}$$

where $C, b > 1$ *and* $\alpha, \kappa > 0$ *are given. If*

$$X_0 + Z_0^{1+\kappa} \leq (2C)^{-\frac{1+\kappa}{\sigma}}\, b^{-\frac{1+\kappa}{\sigma^2}}, \qquad with \quad \sigma = \min\{\alpha, \kappa\},$$

then $X_n, Z_n \to 0$ *as* $n \to \infty$.

3. An Embedding Theorem

Let $V_0^p(\Omega_T)$ denote the space

$$V_0^p(\Omega_T) = L^\infty\left(0, T; L^p(\Omega)\right) \cap L^p\left(0, T; W_0^{1,p}(\Omega)\right)$$

endowed with the norm

$$\|u\|_{V^p(\Omega_T)}^p = \operatorname*{ess\,sup}_{0 \leq t \leq T} \|u(\cdot, t)\|_{p,\Omega}^p + \|\nabla u\|_{p,\Omega_T}^p.$$

The following embedding theorem holds (cf. [14, page 9]).

Theorem 2.11. *Let $p > 1$. There exists a constant γ, depending only on d and p, such that for every $v \in V_0^p(\Omega_T)$,*

$$\|v\|_{p,\Omega_T}^p \leq \gamma \, |\, |v| > 0 \,|^{\frac{p}{d+p}} \, \|v\|_{V^p(\Omega_T)}^p.$$

4. A Poincaré-type Inequality

Let $\Omega \subset \mathbb{R}^d$ be a bounded and convex set. Consider a function $\varphi \in C(\overline{\Omega})$, $0 \leq \varphi \leq 1$, such that the sets

$$[\varphi > k], \quad 0 < k < 1$$

are all convex.
The following theorem holds (cf. [14, page 5]).

Theorem 2.12. *Let $v \in W^{1,p}(\Omega)$, $p \geq 1$ and assume that the set*

$$\Xi := [v = 0] \cap [\varphi = 1]$$

has positive measure. There exists a constant C, depending only on d and p, and independent of v and φ, such that

$$\int_\Omega |v|^p \, \varphi \, dx \leq C \, \frac{|\operatorname{diam} \Omega|^{dp}}{|\Xi|^{\frac{(d-1)p}{d}}} \int_\Omega |\nabla v|^p \, \varphi \, dx.$$

5. Constants

With C we denote constants that depend only on d and p; these constants may be different if they appear in different lines.

3

The Geometric Setting and an Alternative

We go back to equation

$$u_t - \operatorname{div} |\nabla u|^{p-2} \nabla u = 0 \tag{3.1}$$

and focus on the degenerate case $p > 2$. Results on the continuity of solutions at a point consist basically in constructing a sequence of nested and shrinking cylinders with vertex at that point, and in showing that the essential oscillation of the solution in those cylinders converges to zero as the cylinders shrink to the point.

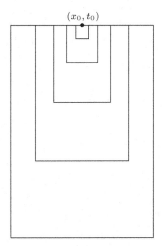

This iterative procedure is based on energy and logarithmic estimates and works well with the standard parabolic cylinders if these estimates are homogeneous. The idea goes back to the work of De Giorgi, Moser and the Russian school (cf. [10], [42] and [37]), as explained in the introduction.

For degenerate or singular equations, the energy and logarithmic estimates are not homogeneous, as we have seen in the previous chapter. They involve integral norms corresponding to different powers, namely the powers 2 and p. To go about this difficulty, the equation has to be analyzed in a geometry dictated by its own degenerate structure. This amounts to rescale the standard parabolic cylinders by a factor that depends on the oscillation of the solution. This procedure of *intrinsic scaling*, which can be seen as an accommodation of the degeneracy, allows for the restoration of the homogeneity in the energy estimates, when written over the rescaled cylinders. We can say heuristically that the equation behaves in its own geometry like the heat equation. Let us make this idea precise.

3.1 A Geometry for the Equation

The standard parabolic cylinders

$$(x_0, t_0) + Q(R^2, R)$$

reflect the natural homogeneity between the space and time variables for the heat equation. Indeed, if $u(x, t)$ is a solution, then $u(\varepsilon x, \varepsilon^2 t)$, $\varepsilon \in \mathbb{R}$, is also a solution, *i.e.*, the equation remains invariant through a similarity transformation of the space-time variables that leaves constant the ratio $|x|^2/t$.

When dealing with the degenerate PDE (3.1), one might think, at first sight, that the adequate cylinders to perform the iterative method described above were cylinders of the form $Q(R^p, R)$, that correspond to the similarity scaling $|x|^p/t$ of the equation. But a more careful analysis shows that this is not to be expected. Indeed, it would work for the homogeneous equation

$$(u^{p-1})_t - \operatorname{div} |\nabla u|^{p-2} \nabla u = 0$$

but not for the inhomogeneous equation (3.1). By analogy, and in order to gain some hindsight on how to proceed, we recast (3.1) in the form

$$\left(\frac{u}{c}\right)^{2-p} (u^{p-1})_t - \operatorname{div} |\nabla u|^{p-2} \nabla u = 0,$$

for an appropriate constant c. This shows that the homogeneity can be recovered at the expense of a scaling factor, that depends on the solution itself and, modulo a constant, looks like u^{2-p}. The following is a sophisticated and rigorous way of implementing this heuristic reasoning.

Consider $0 < R < 1$, sufficiently small so that $Q(R^2, R) \subset \Omega_T$, and define the essential oscillation of the solution u within this cylinder

$$\omega := \operatorname*{ess\,osc}_{Q(R^2, R)} u = \mu^+ - \mu^-,$$

where

$$\mu^+ := \operatorname*{ess\,sup}_{Q(R^2, R)} u \qquad \text{and} \qquad \mu^- := \operatorname*{ess\,inf}_{Q(R^2, R)} u.$$

Then construct the rescaled cylinder

$$Q(a_0 R^p, R) = K_R(0) \times (-a_0 R^p, 0), \quad \text{with} \quad a_0 = \left(\frac{\omega}{2^\lambda}\right)^{2-p}, \tag{3.2}$$

where $\lambda > 1$ is to be fixed later depending only on the data (see (4.15)). Note that for $p = 2$, i.e., in the non-degenerate case, these are the standard parabolic cylinders reflecting the natural homogeneity between the space and time variables.

We will assume, without loss of generality, that

$$R < \frac{\omega}{2^\lambda}. \tag{3.3}$$

Indeed, if this does not hold, we have $\omega \leq 2^\lambda R$ and there is nothing to prove since the oscillation is then comparable to the radius. Now, (3.3) implies the inclusion

$$Q(a_0 R^p, R) \subset Q(R^2, R)$$

and the relation

$$\operatorname*{ess\,osc}_{Q(a_0 R^p, R)} u \leq \omega \tag{3.4}$$

which will be the starting point of an iteration process that leads to the main results. The schematics below give an idea of the stretching procedure, commonly referred to as *accommodation of the degeneracy* (the pictures are distorted on purpose in the t-direction).

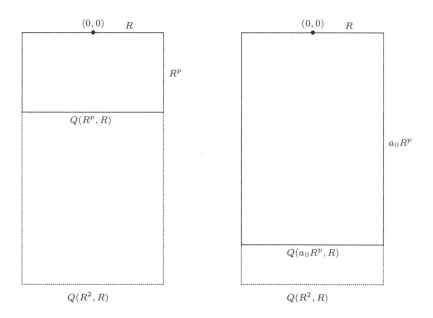

Note that we had to consider the cylinder $Q(R^2, R)$ and assume (3.3), so that (3.4) would hold for the rescaled cylinder $Q(a_0 R^p, R)$. This is in general not true for a given cylinder, since its dimensions would have to be intrinsically defined in terms of the essential oscillation of the function within it.

We now consider subcylinders of $Q(a_0 R^p, R)$ of the form

$$(0, t^*) + Q(\theta R^p, R) , \quad \text{with} \quad \theta = \left(\frac{\omega}{2}\right)^{2-p} \tag{3.5}$$

that are contained in $Q(a_0 R^p, R)$ provided

$$\left(2^{p-2} - 2^{\lambda(p-2)}\right) \frac{R^p}{\omega^{p-2}} < t^* < 0. \tag{3.6}$$

Once λ is chosen, we may redefine it, putting

$$\lambda^* = \frac{[p-1]}{p-2} [\lambda] + 1 > \lambda,$$

and assume that

$$N_0 = \frac{a_0}{\theta} = \left(\frac{\frac{\omega}{2^\lambda}}{\frac{\omega}{2}}\right)^{2-p} = 2^{(\lambda-1)(p-2)} \tag{3.7}$$

is an integer. Thus, we consider $Q(a_0 R^p, R)$ as being divided in subcylinders, all alike and congruent with $Q(\theta R^p, R)$:

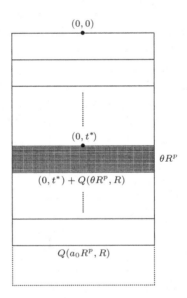

The proof of the Hölder continuity of a weak solution u now follows from the analysis of two complementary cases. We briefly describe them in the following way: in the first case we assume that there is a cylinder of the type $(0, t^*) + Q(\theta R^p, R)$ where u is essentially away from its infimum. We show that going down to a smaller cylinder the oscillation decreases by a small factor that we can exhibit. If that cylinder can not be found then u is essentially away from its supremum in all cylinders of that type and we can compound this information to reach the same conclusion as in the previous case. We state this in a precise way.

For a constant $\nu_0 \in (0, 1)$, that will be determined depending only on the data, either

The First Alternative:

there is a cylinder of the type $(0, t^*) + Q(\theta R^p, R)$ for which

$$\frac{\left|\{(x, t) \in (0, t^*) + Q(\theta R^p, R) \ : \ u(x, t) < \mu^- + \frac{\omega}{2}\}\right|}{|Q(\theta R^p, R)|} \leq \nu_0 \qquad (3.8)$$

or this does not hold. Then, since $\mu^+ - \frac{\omega}{2} = \mu^- + \frac{\omega}{2}$, it holds

The Second Alternative:

for every cylinder of the type $(0, t^*) + Q(\theta R^p, R)$

$$\frac{\left|\{(x, t) \in (0, t^*) + Q(\theta R^p, R) \ : \ u(x, t) > \mu^+ - \frac{\omega}{2}\}\right|}{|Q(\theta R^p, R)|} < 1 - \nu_0. \qquad (3.9)$$

3.2 The First Alternative

We start the analysis assuming the first alternative holds.

Lemma 3.1. *Assume (3.3) is in force. There exists a constant $\nu_0 \in (0, 1)$, depending only on the data, such that if (3.8) holds for some t^* as in (3.6) then*

$$u(x, t) > \mu^- + \frac{\omega}{4}, \quad \text{a.e. in } (0, t^*) + Q\left(\theta \left(\frac{R}{2}\right)^p, \frac{R}{2}\right).$$

Proof. Take the cylinder for which (3.8) holds and assume, by translation, that $t^* = 0$. Let

$$R_n = \frac{R}{2} + \frac{R}{2^{n+1}}, \qquad n = 0, 1, \ldots,$$

and construct the family of nested and shrinking cylinders $Q(\theta R_n^p, R_n)$. Consider piecewise smooth cutoff functions $0 \leq \zeta_n \leq 1$, defined in these cylinders, and satisfying the following set of assumptions:

$$\zeta_n = 1 \quad \text{in } Q\left(\theta R_{n+1}^p, R_{n+1}\right); \qquad \zeta_n = 0 \quad \text{on } \partial_p Q\left(\theta R_n^p, R_n\right);$$

$$|\nabla \zeta_n| \le \frac{2^{n+1}}{R}; \qquad 0 \le (\zeta_n)_t \le \frac{2^{p(n+1)}}{\theta R^p}.$$

Observe that the family of cylinders starts with $Q(\theta R^p, R)$ and converges to $Q\left(\theta \left(\frac{R}{2}\right)^p, \frac{R}{2}\right)$ and that the bounds on the gradient and the time derivative of ζ_n are strictly related to the dimensions of the cylinders.

Write the energy inequality (2.6) over the cylinders $Q\left(\theta R_n^p, R_n\right)$, for the functions $(u - k_n)_-$, with

$$k_n = \mu^- + \frac{\omega}{4} + \frac{\omega}{2^{n+2}}, \quad n = 0, 1, \ldots,$$

and $\zeta = \zeta_n$. They read, taking into account that ζ_n vanishes on $\partial_p Q\left(\theta R_n^p, R_n\right)$,

$$\sup_{-\theta R_n^p < t < 0} \int_{K_{R_n} \times \{t\}} (u - k_n)_-^2 \zeta_n^p \, dx + \int_{-\theta R_n^p}^0 \int_{K_{R_n}} |\nabla(u - k_n)_- \zeta_n|^p \, dx \, dt$$

$$\le C \int_{-\theta R_n^p}^0 \int_{K_{R_n}} (u - k_n)_-^p |\nabla \zeta_n|^p \, dx \, dt + p \int_{-\theta R_n^p}^0 \int_{K_{R_n}} (u - k_n)_-^2 \zeta_n^{p-1} (\zeta_n)_t \, dx \, dt$$

$$\le C \frac{2^{p(n+1)}}{R^p} \left\{ \int_{-\theta R_n^p}^0 \int_{K_{R_n}} (u - k_n)_-^p \, dx \, dt + \frac{1}{\theta} \int_{-\theta R_n^p}^0 \int_{K_{R_n}} (u - k_n)_-^2 \, dx \, dt \right\}.$$

Next, observe that either $(u - k_n)_- = 0$ or

$$(u - k_n)_- = (\mu^- - u) + \frac{\omega}{4} + \frac{\omega}{2^{n+2}} \le \frac{\omega}{2},$$

and thus, since $2 - p < 0$,

$$(u - k_n)_-^2 = (u - k_n)_-^{2-p} (u - k_n)_-^p$$
$$\ge \left(\frac{\omega}{2}\right)^{2-p} (u - k_n)_-^p$$
$$= \theta (u - k_n)_-^p,$$

recalling that $\theta = \left(\frac{\omega}{2}\right)^{2-p}$. We obtain, homogenizing the powers in the integral norms,

$$\theta \sup_{-\theta R_n^p < t < 0} \int_{K_{R_n} \times \{t\}} (u - k_n)_-^p \zeta_n^p \, dx + \int_{-\theta R_n^p}^0 \int_{K_{R_n}} |\nabla(u - k_n)_- \zeta_n|^p \, dx \, dt$$

$$\le C \frac{2^{p(n+1)}}{R^p} \left\{ \left(\frac{\omega}{2}\right)^p + \frac{1}{\theta} \left(\frac{\omega}{2}\right)^2 \right\} \int_{-\theta R_n^p}^0 \int_{K_{R_n}} \chi_{\{(u-k_n)_->0\}} \, dx \, dt,$$

where χ_E denotes the characteristic function of the set E. Finally, divide throughout by θ to get

$$\sup_{-\theta R_n^p < t < 0} \int_{K_{R_n} \times \{t\}} (u - k_n)_-^p \zeta_n^p \, dx + \frac{1}{\theta} \int_{-\theta R_n^p}^0 \int_{K_{R_n}} |\nabla (u - k_n)_- \zeta_n|^p \, dx \, dt$$

$$\le C \frac{2^{p(n+1)}}{R^p} \left(\frac{\omega}{2}\right)^p \frac{1}{\theta} \int_{-\theta R_n^p}^0 \int_{K_{R_n}} \chi_{\{(u-k_n)_->0\}} \, dx \, dt. \qquad (3.10)$$

The next step, in which the intrinsic geometric framework is crucial, is to perform a change in the time variable, putting $\bar{t} = t/\theta$, and to define

$$\bar{u}(\cdot, \bar{t}) := u(\cdot, t) , \qquad \overline{\zeta_n}(\cdot, \bar{t}) := \zeta_n(\cdot, t).$$

We obtain the simplified inequality

$$\left\|(\bar{u} - k_n)_- \overline{\zeta_n}\right\|_{V^p(Q(R_n^p, R_n))}^p \le C \frac{2^{pn}}{R^p} \left(\frac{\omega}{2}\right)^p \int_{-R_n^p}^0 \int_{K_{R_n}} \chi_{\{(\bar{u}-k_n)_->0\}} \, dx \, d\bar{t},$$
$$(3.11)$$

which reveals the appropriate functional framework.

To conclude, define, for each n,

$$A_n = \int_{-R_n^p}^0 \int_{K_{R_n}} \chi_{\{(\bar{u}-k_n)_->0\}} \, dx \, d\bar{t}$$

and observe that

$$\frac{1}{2^{p(n+2)}} \left(\frac{\omega}{2}\right)^p A_{n+1} = |k_n - k_{n+1}|^p \, A_{n+1}$$

$$\le \left\|(\bar{u} - k_n)_-\right\|_{p,Q(R_{n+1}^p, R_{n+1})}^p$$

$$\le \left\|(\bar{u} - k_n)_- \overline{\zeta_n}\right\|_{p,Q(R_n^p, R_n)}^p$$

$$\le C \left\|(\bar{u} - k_n)_- \overline{\zeta_n}\right\|_{V^p(Q(R_n^p, R_n))}^p A_n^{\frac{p}{d+p}}$$

$$\le C \frac{2^{pn}}{R^p} \left(\frac{\omega}{2}\right)^p A_n^{1+\frac{p}{d+p}}. \qquad (3.12)$$

[The first two inequalities follow from the definition of A_n and the fact that $k_{n+1} < k_n$; the third inequality is a consequence of Theorem 2.11 and the last one follows from (3.11).] Next, define the numbers

$$X_n = \frac{A_n}{|Q(R_n^p, R_n)|},$$

divide (3.12) by $|Q(R_{n+1}^p, R_{n+1})|$ and obtain the recursive relation

$$X_{n+1} \le C \, 4^{pn} \, X_n^{1+\frac{p}{d+p}},$$

for a constant C depending only upon d and p. By Lemma 2.9 on fast geometric convergence, if

$$X_0 \le C^{-\frac{d+p}{p}} 4^{-\frac{(d+p)^2}{p}} =: \nu_0 \qquad (3.13)$$

then

$$X_n \longrightarrow 0. \tag{3.14}$$

But (3.13) is precisely our hypothesis (3.8), for the indicated choice of ν_0, and from (3.14) we immediately obtain, returning to the original variables,

$$\left| \left\{ (x,t) \in Q\left(\theta(\tfrac{R}{2})^p, \tfrac{R}{2}\right) \; : \; u(x,t) \le \mu^- + \frac{\omega}{4} \right\} \right| = 0.$$

□

Remark 3.2. The constant ν_0, that appears in the formulation of the alternative, is now fixed by (3.13). Note that indeed $\nu_0 \in (0,1)$.

3.3 The Role of the Logarithmic Estimates: Expansion in Time

Our next aim is to show that the conclusion of Lemma 3.1 holds in a full cylinder $Q(\tau, \rho)$. The idea is to use the fact that at the time level

$$-\widehat{t} := t^* - \theta\left(\tfrac{R}{2}\right)^p \tag{3.15}$$

the function $u(x, -\widehat{t}\,)$ is strictly above the level $\mu^- + \frac{\omega}{4}$ in the cube $K_{\frac{R}{2}}$, and look at this time level as an initial condition to make the conclusion hold up to $t = 0$ in a smaller cylinder, as sketched in the following diagram:

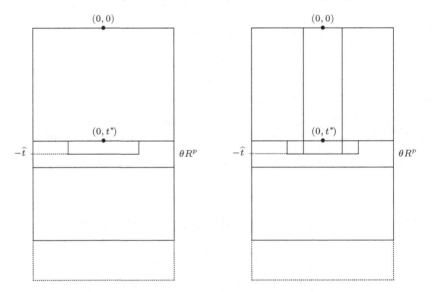

As an intermediate step we need the following lemma, in which the use of the logarithmic estimates is crucial.

Lemma 3.3. *Assume (3.8) holds for some t^* as in (3.6) and that (3.3) is in force. Given $\nu_* \in (0,1)$, there exists $s_* \in \mathbb{N}$, depending only on the data, such that*

$$\left| \left\{ x \in K_{\frac{R}{4}} \; : \; u(x,t) < \mu^- + \frac{\omega}{2^{s_*}} \right\} \right| \leq \nu_* \left| K_{\frac{R}{4}} \right|, \quad \forall t \in (-\hat{t}, 0).$$

Proof. We use the logarithmic estimate (2.8) applied to the function $(u-k)_-$ in the cylinder $Q(\hat{t}, \frac{R}{2})$, with the choices

$$k = \mu^- + \frac{\omega}{4} \quad \text{and} \quad c = \frac{\omega}{2^{n+2}},$$

where $n \in \mathbb{N}$ will be chosen later. In this cylinder, we have

$$k - u \leq H^-_{u,k} = \operatorname*{ess\,sup}_{Q(\hat{t}, \frac{R}{2})} \left| \left(u - \mu^- - \frac{\omega}{4} \right)_- \right| \leq \frac{\omega}{4}. \tag{3.16}$$

If $H^-_{u,k} \leq \frac{\omega}{8}$, the result is trivial for the choice $s^* = 3$. Assuming $H^-_{u,k} > \frac{\omega}{8}$, recall from section 2.3 that the logarithmic function $\psi^-(u)$ is defined in the whole of $Q(\hat{t}, \frac{R}{2})$ and it is given by

$$\psi^-_{\{H^-_{u,k}, k, \frac{\omega}{2^{n+2}}\}}(u) = \begin{cases} \ln \left\{ \dfrac{H^-_{u,k}}{H^-_{u,k} + u - k + \frac{\omega}{2^{n+2}}} \right\} & \text{if } u < k - \frac{\omega}{2^{n+2}} \\[4mm] 0 & \text{if } u \geq k - \frac{\omega}{2^{n+2}}. \end{cases}$$

From (3.16), we estimate

$$\psi^-(u) \leq n \ln 2 \quad \text{since} \quad \frac{H^-_{u,k}}{H^-_{u,k} + u - k + \frac{\omega}{2^{n+2}}} \leq \frac{\frac{\omega}{4}}{\frac{\omega}{2^{n+2}}} = 2^n \tag{3.17}$$

and

$$\left| (\psi^-)'(u) \right|^{2-p} = \left(H^-_{u,k} + u - k + c \right)^{p-2} \leq \left(\frac{\omega}{2} \right)^{p-2}. \tag{3.18}$$

Now observe that, as a consequence of Lemma 3.1, we have $u(x, -\hat{t}) > k$ in the cube $K_{\frac{R}{2}}$, which implies that

$$\left[\psi^-(u) \right](x, -\hat{t}) = 0, \quad x \in K_{\frac{R}{2}}.$$

Choosing a piecewise smooth cutoff function $0 < \zeta(x) \leq 1$, defined in $K_{\frac{R}{2}}$ and such that

$$\zeta = 1 \quad \text{in } K_{\frac{R}{4}} \quad \text{and} \quad |\nabla \zeta| \leq \frac{8}{R},$$

inequality (2.8) reads

$$\sup_{-\hat{t}<t<0} \int_{K_{\frac{R}{2}}\times\{t\}} \left[\psi^-(u)\right]^2 \zeta^p \, dx$$

$$\leq C \int_{-\hat{t}}^0 \int_{K_{\frac{R}{2}}} \psi^-(u) \left|(\psi^-)'(u)\right|^{2-p} |\nabla\zeta|^p \, dx \, dt. \qquad (3.19)$$

The right hand side is estimated above, using (3.17) and (3.18), by

$$C \, n (\ln 2) \left(\frac{\omega}{2}\right)^{p-2} \left(\frac{8}{R}\right)^p \hat{t} \left|K_{\frac{R}{2}}\right| \leq C \, n \, 2^{\lambda(p-2)} \left|K_{\frac{R}{4}}\right|,$$

since, by (3.15),

$$\hat{t} \leq a_0 R^p = \left(\frac{\omega}{2^\lambda}\right)^{2-p} R^p.$$

We estimate below the left hand side of (3.19) by integrating over the smaller set

$$S = \left\{x \in K_{\frac{R}{4}} : u(x,t) < \mu^- + \frac{\omega}{2^{n+2}}\right\} \subset K_{\frac{R}{2}},$$

and observing that in S, $\zeta = 1$ and

$$\frac{H_{u,k}^-}{H_{u,k}^- + u - k + \frac{\omega}{2^{n+2}}}$$

is a decreasing function of $H_{u,k}^-$ because $u - k + \frac{\omega}{2^{n+2}} < 0$. Thus, from (3.16),

$$\frac{H_{u,k}^-}{H_{u,k}^- + u - k + \frac{\omega}{2^{n+2}}} \geq \frac{\frac{\omega}{4}}{\frac{\omega}{4} + u - k + \frac{\omega}{2^{n+2}}} = \frac{\frac{\omega}{4}}{u - \mu^- + \frac{\omega}{2^{n+2}}} > \frac{\frac{\omega}{4}}{\frac{\omega}{2^{n+1}}} = 2^{n-1}$$

since $u - \mu^- < \frac{\omega}{2^{n+2}}$ in S. Therefore,

$$\left[\psi^-(u)\right]^2 \geq \left[\ln\left(2^{n-1}\right)\right]^2 = (n-1)^2 (\ln 2)^2 \quad \text{in } S.$$

Combining these estimates in (3.19), we get

$$\left|\left\{x \in K_{\frac{R}{4}} : u(x,t) < \mu^- + \frac{\omega}{2^{n+2}}\right\}\right| \leq C \, \frac{n}{(n-1)^2} \, 2^{\lambda(p-2)} \left|K_{\frac{R}{4}}\right|,$$

for all $t \in (-\hat{t}, 0)$, and to prove the lemma we choose

$$s_* = n + 2 \qquad \text{with} \qquad n > 1 + \frac{2C}{\nu_*} \, 2^{\lambda(p-2)}.$$

$\square$

3.4 Reduction of the Oscillation

We now state the main result in the context of the first alternative.

Proposition 3.4. *Assume (3.8) holds for some t^* as in (3.6) and that (3.3) is in force. There exists $s_1 \in \mathbb{N}$, depending only on the data, such that*

$$u(x,t) > \mu^- + \frac{\omega}{2^{s_1+1}}, \quad a.e. \ in \ Q\left(\hat{t}, \frac{R}{8}\right).$$

Proof. Consider the cylinder for which (3.8) holds, let

$$R_n = \frac{R}{8} + \frac{R}{2^{n+3}}, \quad n = 0, 1, \ldots$$

and construct the family of nested and shrinking cylinders $Q(\hat{t}, R_n)$, where $\hat{t}$ is given by (3.15). Take piecewise smooth cutoff functions $0 < \zeta_n(x) \leq 1$, independent of t, defined in K_{R_n} and satisfying

$$\zeta_n = 1 \quad \text{in} \ K_{R_{n+1}}; \qquad |\nabla \zeta_n| \leq \frac{2^{n+4}}{R}.$$

Write the local energy inequalities (2.6) for the functions $(u - k_n)_-$, in the cylinders $Q(\hat{t}, R_n)$, with

$$k_n = \mu^- + \frac{\omega}{2^{s_1+1}} + \frac{\omega}{2^{s_1+1+n}}, \quad n = 0, 1, \ldots,$$

s_1 to be chosen, and $\zeta = \zeta_n$. Observing that, due to Lemma 3.1, we have

$$u(x, -\hat{t}) > \mu^- + \frac{\omega}{4} \geq k_n \quad \text{in} \ K_{\frac{R}{2}} \supset K_{R_n},$$

which implies that

$$(u - k_n)_-(x, -\hat{t}) = 0 \quad \text{in} \ K_{R_n}, \quad n = 0, 1, \ldots,$$

the estimates read

$$\sup_{-\hat{t} < t < 0} \int_{K_{R_n} \times \{t\}} (u - k_n)_-^2 \, \zeta_n^p \, dx + \int_{-\hat{t}}^0 \int_{K_{R_n}} |\nabla(u - k_n)_- \zeta_n|^p \, dx \, dt$$

$$\leq C \int_{-\hat{t}}^0 \int_{K_{R_n}} (u - k_n)_-^p |\nabla \zeta_n|^p \, dx \, dt$$

$$\leq C \frac{2^{p(n+4)}}{R^p} \int_{-\hat{t}}^0 \int_{K_{R_n}} (u - k_n)_-^p \, dx \, dt. \tag{3.20}$$

From (3.15), we estimate

$$\hat{t} \leq a_0 R^p = \left(\frac{\omega}{2^\lambda}\right)^{2-p} R^p,$$

where a_0 is defined in (3.2). From this,

$$
\begin{aligned}
(u - k_n)_-^2 &\geq \left(\frac{\omega}{2^{s_1}}\right)^{2-p} (u - k_n)_-^p \\
&\geq \left(\frac{2^{s_1}}{2^\lambda}\right)^{p-2} \frac{\widehat{t}}{R^p} (u - k_n)_-^p \\
&\geq \frac{\widehat{t}}{(\frac{R}{2})^p} (u - k_n)_-^p,
\end{aligned}
$$

provided $s_1 > \lambda + \frac{p}{p-2}$. Dividing now by $\frac{\widehat{t}}{(\frac{R}{2})^p}$ throughout (3.20) gives

$$
\sup_{-\widehat{t} < t < 0} \int_{K_{R_n} \times \{t\}} (u - k_n)_-^p \zeta_n^p \, dx + \frac{(\frac{R}{2})^p}{\widehat{t}} \int_{-\widehat{t}}^0 \int_{K_{R_n}} |\nabla (u - k_n)_- \zeta_n|^p \, dx \, dt
$$
$$
\leq C \frac{2^{pn}}{\widehat{t}} \int_{-\widehat{t}}^0 \int_{K_{R_n}} (u - k_n)_-^p \, dx \, dt.
$$

The change of the time variable $\overline{t} = t \frac{(\frac{R}{2})^p}{\widehat{t}}$, along with defining the new function

$$
\overline{u}(\cdot, \overline{t}) := u(\cdot, t),
$$

leads to the simplified inequality

$$
\|(\overline{u} - k_n)_- \zeta_n\|_{V^p(Q((\frac{R}{2})^p, R_n))}^p \leq C \frac{2^{pn}}{(\frac{R}{2})^p} \left(\frac{\omega}{2^{s_1}}\right)^p \int_{-(\frac{R}{2})^p}^0 \int_{K_{R_n}} \chi_{\{\overline{u} < k_n\}} \, dx \, d\overline{t}.
$$

Define, for each n,

$$
A_n = \int_{-(\frac{R}{2})^p}^0 \int_{K_{R_n}} \chi_{\{(\overline{u} - k_n)_- > 0\}} \, dx \, d\overline{t}.
$$

By a reasoning similar to the one leading to (3.12), we obtain

$$
\begin{aligned}
\frac{1}{2^{p(n+2)}} \left(\frac{\omega}{2^{s_1}}\right)^p A_{n+1} &= |k_n - k_{n+1}|^p A_{n+1} \\
&\leq \|(\overline{u} - k_n)_-\|_{p, Q((\frac{R}{2})^p, R_{n+1})}^p \\
&\leq \|(\overline{u} - k_n)_- \zeta_n\|_{p, Q((\frac{R}{2})^p, R_n)}^p \\
&\leq C \|(\overline{u} - k_n)_- \zeta_n\|_{V^p(Q((\frac{R}{2})^p, R_n))}^p A_n^{\frac{p}{d+p}} \\
&\leq C \frac{2^{pn}}{(\frac{R}{2})^p} \left(\frac{\omega}{2^{s_1}}\right)^p A_n^{1 + \frac{p}{d+p}}.
\end{aligned}
$$

Next, define the numbers

$$X_n = \frac{A_n}{\left| Q\left((\tfrac{R}{2})^p, R_n \right) \right|},$$

and divide the previous inequality by $\left| Q\left((\tfrac{R}{2})^p, R_{n+1} \right) \right|$ to obtain the recursive relations

$$X_{n+1} \leq C \, 4^{pn} \, X_n^{1+\frac{p}{d+p}}.$$

By Lemma 2.9 on fast geometric convergence, if

$$X_0 \leq C^{-\frac{d+p}{p}} 4^{-\frac{(d+p)^2}{p}} =: \nu_* \in (0,1) \tag{3.21}$$

then

$$X_n \longrightarrow 0. \tag{3.22}$$

Apply Lemma 3.3 with such a ν_* and conclude that there exists $s_* =: s_1$, depending only on the data, such that

$$\left| \left\{ x \in K_{\frac{R}{4}} \; : \; u(x,t) < \mu^- + \frac{\omega}{2^{s_1}} \right\} \right| \leq \nu_* \left| K_{\frac{R}{4}} \right|, \quad \forall t \in (-\hat{t}, 0),$$

which is exactly (3.21). Since (3.22) implies that $A_n \to 0$, we conclude that

$$\left| \left\{ (x, \bar{t}) \in Q\left((\tfrac{R}{2})^p, \frac{R}{8} \right) \; : \; \bar{u}(x, \bar{t}) \leq \mu^- + \frac{\omega}{2^{s_1+1}} \right\} \right|$$
$$= \left| \left\{ (x, t) \in Q\left(\hat{t}, \frac{R}{8} \right) \; : \; u(x, t) \leq \mu^- + \frac{\omega}{2^{s_1+1}} \right\} \right| = 0.$$

$\square$

We finally reach the conclusion of the first alternative, namely the reduction of the oscillation.

Corollary 3.5. *Assume (3.8) holds for some t^* as in (3.6) and that (3.3) is in force. There exists a constant $\sigma_0 \in (0,1)$, depending only on the data, such that*

$$\operatorname*{ess\,osc}_{Q\left(\theta(\frac{R}{8})^p, \frac{R}{8}\right)} u \leq \sigma_0 \, \omega. \tag{3.23}$$

Proof. By Proposition 3.4, there exists $s_1 \in \mathbb{N}$ such that

$$\operatorname*{ess\,inf}_{Q(\hat{t}, \frac{R}{8})} u \geq \mu^- + \frac{\omega}{2^{s_1+p}}$$

and thus

$$\operatorname*{ess\,osc}_{Q(\hat{t},\frac{R}{8})} u = \operatorname*{ess\,sup}_{Q(\hat{t},\frac{R}{8})} u - \operatorname*{ess\,inf}_{Q(\hat{t},\frac{R}{8})} u$$

$$\leq \mu^{+} - \mu^{-} - \frac{\omega}{2^{s_1+1}}$$

$$= \left(1 - \frac{1}{2^{s_1+1}}\right)\omega.$$

Since $\theta\left(\frac{R}{8}\right)^{p} \leq \hat{t} = -t^* + \theta\left(\frac{R}{2}\right)^{p}$, with $t^* < 0$, we have

$$Q\left(\theta\left(\frac{R}{8}\right)^{p}, \frac{R}{8}\right) \subset Q\left(\hat{t}, \frac{R}{8}\right),$$

and the corollary follows with $\sigma_0 = \left(1 - \frac{1}{2^{s_1+1}}\right)$. $\square$

4

Towards the Hölder Continuity

Assume now that the second alternative (3.9) holds true: for every cylinder of the type $(0, t^*) + Q(\theta R^p, R)$

$$\frac{\left|\left\{(x,t) \in (0,t^*) + Q(\theta R^p, R) \ : \ u(x,t) > \mu^+ - \frac{\omega}{2}\right\}\right|}{|Q(\theta R^p, R)|} < 1 - \nu_0. \qquad (4.1)$$

This is somehow the unfavorable case but we will show that a conclusion similar to (3.23) can still be reached. The idea is to exploit the fact that (4.1) holds for all cylinders of the type mentioned and add up that information to obtain the desired conclusion. Recall that the constant ν_0 has already been quantitatively determined by (3.13), and it is now fixed. We continue to assume that (3.3) is in force.

4.1 Expanding in Time

Fix a cylinder $(0, t^*) + Q(\theta R^p, R) \subset Q(a_0 R^p, R)$ for which (4.1) holds. Then there exists a time level

$$t^\circ \in \left[t^* - \theta R^p, t^* - \frac{\nu_0}{2}\theta R^p\right]$$

such that

$$\left|\left\{x \in K_R \ : \ u(x,t^\circ) > \mu^+ - \frac{\omega}{2}\right\}\right| \leq \left(\frac{1 - \nu_0}{1 - \nu_0/2}\right)|K_R|. \qquad (4.2)$$

Otherwise, for all $t \in \left[t^* - \theta R^p, t^* - \frac{\nu_0}{2}\theta R^p\right]$, we would have

$$\left|\left\{(x,t) \in (0,t^*) + Q(\theta R^p, R) \ : \ u(x,t) > \mu^+ - \frac{\omega}{2}\right\}\right|$$

$$\geq \int_{t^* - \theta R^p}^{t^* - \frac{\nu_0}{2}\theta R^p} \left|\left\{x \in K_R \ : \ u(x,\tau) > \mu^+ - \frac{\omega}{2}\right\}\right| d\tau$$

$$> (1 - \nu_0)\,|Q(\theta R^p, R)|,$$

which contradicts (4.1).

The next lemma asserts that the set where $u(\cdot, t)$ is close to its supremum is small, not only at the specific time level $t°$, but for all time levels near the top of the cylinder $(0, t^*) + Q(\theta R^p, R)$.

Lemma 4.1. *Assume (4.1) and let (3.3) be in force. There exists $s_2 \in \mathbb{N}$, depending only on the data, such that*

$$\left| \left\{ x \in K_R : u(x,t) > \mu^+ - \frac{\omega}{2^{s_2}} \right\} \right| \leq \left(1 - \left(\frac{\nu_0}{2} \right)^2 \right) |K_R|,$$

for all $t \in \left[t^* - \frac{\nu_0}{2} \theta R^p, t^* \right]$.

Proof. The proof consists in using the logarithmic inequalities (2.8) applied to the function $(u-k)_+$ in the cylinder $K_R \times (t°, t^*)$, with the choices

$$k = \mu^+ - \frac{\omega}{2} \qquad \text{and} \qquad c = \frac{\omega}{2^{n+1}},$$

where $n \in \mathbb{N}$ will be chosen later. In this cylinder, we have

$$u - k \leq H_{u,k}^+ = \operatorname*{ess\,sup}_{K_R \times (t°,t^*)} \left| \left(u - \mu^+ + \frac{\omega}{2} \right)_+ \right| \leq \frac{\omega}{2}. \tag{4.3}$$

If $H_{u,k}^+ \leq \frac{\omega}{4}$ the result is trivial with the choice $s_2 = 2$. Assuming $H_{u,k}^+ > \frac{\omega}{4}$, recall from section 2.3 that the logarithmic function $\psi^+(u)$ is defined in the whole of $K_R \times (t°, t^*)$, and it is given by

$$\psi^+_{\{H_{u,k}^+, k, \frac{\omega}{2^{n+1}}\}}(u) = \begin{cases} \ln \left\{ \dfrac{H_{u,k}^+}{H_{u,k}^+ - u + k + \frac{\omega}{2^{n+1}}} \right\} & \text{if } u > k + \frac{\omega}{2^{n+1}} \\[2ex] 0 & \text{if } u \leq k + \frac{\omega}{2^{n+1}}. \end{cases}$$

From (4.3), we obtain the estimates

$$\psi^+(u) \leq n \ln 2 \qquad \text{since} \qquad \frac{H_{u,k}^+}{H_{u,k}^+ - u + k + \frac{\omega}{2^{n+1}}} \leq \frac{\frac{\omega}{2}}{\frac{\omega}{2^{n+1}}} = 2^n, \tag{4.4}$$

and

$$\left| (\psi^+)'(u) \right|^{2-p} = \left(H_{u,k}^+ - u + k + c \right)^{p-2} \leq \left(\frac{\omega}{2} \right)^{p-2}. \tag{4.5}$$

Choosing a piecewise smooth cutoff function $0 < \zeta(x) \leq 1$, defined in K_R and such that, for some $\sigma \in (0, 1)$,

$$\zeta = 1 \text{ in } K_{(1-\sigma)R} \qquad \text{and} \qquad |\nabla \zeta| \leq (\sigma R)^{-1},$$

inequality (2.8) reads

$$\sup_{t^\circ < t < t^*} \int_{K_R \times \{t\}} \left[\psi^+(u)\right]^2 \zeta^p \, dx \leq \int_{K_R \times \{t^\circ\}} \left[\psi^+(u)\right]^2 \zeta^p \, dx$$

$$+ C \int_{t^\circ}^{t^*} \int_{K_R} \psi^+(u) \left|(\psi^+)'(u)\right|^{2-p} |\nabla \zeta|^p \, dx \, dt. \tag{4.6}$$

The first integral on the right hand side can be bounded above using (4.2) and taking into account that $\psi^+(u)$ vanishes on the set

$$\{x \in K_R : u(x, \cdot) \leq k\}.$$

This gives, using also (4.4),

$$\int_{K_R \times \{t^\circ\}} \left[\psi^+(u)\right]^2 \zeta^p \, dx \leq n^2 (\ln 2)^2 \left(\frac{1 - \nu_0}{1 - \nu_0/2}\right) |K_R|.$$

To bound the second integral we use (4.4) and (4.5):

$$C \int_{t^\circ}^{t^*} \int_{K_R} \psi^+(u) \left|(\psi^+)'(u)\right|^{2-p} |\nabla \zeta|^p \, dx \, dt$$

$$\leq C \, n (\ln 2) \left(\frac{\omega}{2}\right)^{p-2} (\sigma R)^{-p} (t^* - t^\circ) |K_R|$$

$$\leq C n \left(\frac{\omega}{2}\right)^{p-2} \left(\frac{1}{\sigma R}\right)^p \theta R^p |K_R|$$

$$\leq C n \frac{1}{\sigma^p} |K_R|,$$

since $t^* - t^\circ \leq \theta R^p$ and $\theta = \left(\frac{\omega}{2}\right)^{2-p}$.

The left hand side is estimated below by integrating over the smaller set

$$S = \left\{x \in K_{(1-\sigma)R} : u(x, t) > \mu^+ - \frac{\omega}{2^{n+1}}\right\} \subset K_R$$

and observing that in S, $\zeta = 1$ and

$$\frac{H_{u,k}^+}{H_{u,k}^+ - u + k + \frac{\omega}{2^{n+1}}}$$

is a decreasing function of $H_{u,k}^+$ because $-u + k + \frac{\omega}{2^{n+1}} < 0$. Thus, from (4.3),

$$\frac{H_{u,k}^+}{H_{u,k}^+ - u + k + \frac{\omega}{2^{n+1}}} \geq \frac{\frac{\omega}{2}}{\frac{\omega}{2} - u + k + \frac{\omega}{2^{n+1}}} = \frac{\frac{\omega}{2}}{-u + \mu^+ + \frac{\omega}{2^{n+1}}} > \frac{\frac{\omega}{2}}{\frac{\omega}{2^n}} = 2^{n-1}$$

since $-u + \mu^+ < \frac{\omega}{2^{n+1}}$ in S. Therefore

$$\left[\psi^+(u)\right]^2 \geq \left[\ln \left(2^{n-1}\right)\right]^2 = (n-1)^2 (\ln 2)^2 \qquad \text{in } S$$

and, from this,

$$\sup_{t^\circ < t < t^*} \int_{K_R \times \{t\}} \left[\psi^+(u)\right]^2 \zeta^p \, dx \geq (n-1)^2 (\ln 2)^2 \, |S|.$$

Combining the three estimates, we arrive at

$$|S| \leq \left(\frac{n}{n-1}\right)^2 \left(\frac{1-\nu_0}{1-\nu_0/2}\right) |K_R| + C \frac{n}{(n-1)^2} \frac{1}{\sigma^p} |K_R|$$

$$\leq \left(\frac{n}{n-1}\right)^2 \left(\frac{1-\nu_0}{1-\nu_0/2}\right) |K_R| + \frac{C}{n\sigma^p} |K_R|.$$

On the other hand,

$$\left|\left\{x \in K_R \, : \, u(x,t) > \mu^+ - \frac{\omega}{2^{n+1}}\right\}\right|$$

$$\leq \left|\left\{x \in K_{(1-\sigma)R} \, : \, u(x,t) > \mu^+ - \frac{\omega}{2^{n+1}}\right\}\right| + \left|K_R \setminus K_{(1-\sigma)R}\right|$$

$$\leq |S| + d\sigma \, |K_R|,$$

and thus

$$\left|\left\{x \in K_R \, : \, u(x,t) > \mu^+ - \frac{\omega}{2^{n+1}}\right\}\right|$$

$$\leq \left\{\left(\frac{n}{n-1}\right)^2 \left(\frac{1-\nu_0}{1-\nu_0/2}\right) + \frac{C}{n\sigma^p} + d\sigma\right\} |K_R|,$$

for all $t \in (t^\circ, t^*)$. Choose σ so small that $d\sigma \leq \frac{3}{8}\nu_0^2$ and n so large that

$$\left(\frac{n}{n-1}\right)^2 \leq \left(1 - \frac{\nu_0}{2}\right)(1+\nu_0) =: \beta \quad \text{and} \quad \frac{C}{n\sigma^p} \leq \frac{3}{8}\nu_0^2. \tag{4.7}$$

Note that $\beta > 1$. With this choice of n, the lemma follows with $s_2 = n+1$. $\square$

The same type of conclusion holds in an upper portion of the full cylinder $Q(a_0 R^p, R)$, say for all $t \in \left(-\frac{a_0}{2}R^p, 0\right)$. Indeed, (4.1) holds for *all* cylinders of the type $(0, t^*) + Q(\theta R^p, R)$ so the conclusion of the previous lemma holds true for all time levels

$$t \geq -(a_0 - \theta)R^p - \frac{\nu_0}{2}\theta R^p.$$

So if we choose λ such that

$$2^{(\lambda-1)(p-2)} \geq 2 \tag{4.8}$$

we get, recalling (3.7), $\frac{a_0}{\theta} \geq 2 - \nu_0$, which is equivalent to

$$-(a_0 - \theta)R^p - \frac{\nu_0}{2}\theta R^p \leq -\frac{a_0}{2}R^p.$$

Corollary 4.2. *Assume (4.1) and let (3.3) be in force. Then*

$$\left| \left\{ x \in K_R : u(x,t) > \mu^+ - \frac{\omega}{2^{s_2}} \right\} \right| \le \left(1 - \left(\frac{\nu_0}{2} \right)^2 \right) |K_R|,$$

for all $t \in \left(-\frac{a_0}{2} R^p, 0 \right)$.

4.2 Reducing the Oscillation

The main result of this chapter states that in fact u is strictly below its supremum μ^+ in a smaller cylinder with the same vertex and axis as $Q\left(\frac{a_0}{2} R^p, R\right)$.

Proposition 4.3. *Assume (4.1) and let (3.3) be in force. The choice of λ can be made so that*

$$u(x,t) \le \mu^+ - \frac{\omega}{2^{\lambda+1}}, \quad a.e. \ in \ Q\left(\frac{a_0}{2} \left(\frac{R}{2} \right)^p, \frac{R}{2} \right). \tag{4.9}$$

Proof. Define

$$R_n = \frac{R}{2} + \frac{R}{2^{n+1}}, \quad n = 0, 1, \dots,$$

and construct the family of nested and shrinking cylinders $Q\left(\frac{a_0}{2} R_n^p, R_n\right)$. Consider piecewise smooth cutoff functions $0 \le \zeta_n \le 1$, defined in these cylinders and satisfying the following set of assumptions:

$$\zeta_n = 1 \quad \text{in } Q\left(\frac{a_0}{2} R_{n+1}^p, R_{n+1}\right); \qquad \zeta_n = 0 \quad \text{on } \partial_p Q\left(\frac{a_0}{2} R_n^p, R_n\right);$$

$$|\nabla \zeta_n| \le \frac{2^{n+1}}{R}; \qquad\qquad 0 \le (\zeta_n)_t \le \frac{2^{p(n+1)}}{\frac{a_0}{2} R^p}.$$

The energy inequality (2.6) for the functions $(u - k_n)_+$, with

$$k_n = \mu^+ - \frac{\omega}{2^{\lambda+1}} - \frac{\omega}{2^{\lambda+1+n}}, \quad n = 0, 1, \dots,$$

in the cylinders $Q\left(\frac{a_0}{2} R_n^p, R_n\right)$, and with $\zeta = \zeta_n$, reads

$$\sup_{-\frac{a_0}{2} R_n^p < t < 0} \int_{K_{R_n} \times \{t\}} (u - k_n)_+^2 \zeta_n^p \, dx + \int_{-\frac{a_0}{2} R_n^p}^{0} \int_{K_{R_n}} |\nabla (u - k_n)_+ \zeta_n|^p \, dx \, dt$$

$$\le C \int_{-\frac{a_0}{2} R_n^p}^{0} \int_{K_{R_n}} (u - k_n)_+^p |\nabla \zeta_n|^p \, dx \, dt$$

$$+ C \int_{-\frac{a_0}{2} R_n^p}^{0} \int_{K_{R_n}} (u - k_n)_+^2 \zeta_n^{p-1} (\zeta_n)_t \, dx \, dt$$

$$\le C \frac{2^{p(n+1)}}{R^p} \left\{ \int_{-\frac{a_0}{2} R_n^p}^{0} \int_{K_{R_n}} (u - k_n)_+^p \, dx \, dt \right.$$

$$\left. + \frac{2}{a_0} \int_{-\frac{a_0}{2} R_n^p}^{0} \int_{K_{R_n}} (u - k_n)_+^2 \, dx \, dt \right\}. \tag{4.10}$$

Observe that

$$(u - k_n)_+^2 = (u - k_n)_+^{2-p} (u - k_n)_+^p \geq \left(\frac{\omega}{2^\lambda}\right)^{2-p} (u - k_n)_+^p.$$

Therefore, from (4.10),

$$\left(\frac{\omega}{2^\lambda}\right)^{2-p} \sup_{-\frac{a_0}{2} R_n^p < t < 0} \int_{K_{R_n} \times \{t\}} (u - k_n)_+^p \zeta_n^p \, dx$$

$$+ \int_{-\frac{a_0}{2} R_n^p}^0 \int_{K_{R_n}} |\nabla(u - k_n)_+ \zeta_n|^p \, dx \, dt$$

$$\leq C \frac{2^{p(n+1)}}{R^p} \left\{ \left(\frac{\omega}{2^\lambda}\right)^p + \frac{2}{a_0} \left(\frac{\omega}{2^\lambda}\right)^2 \right\} \int_{-\frac{a_0}{2} R_n^p}^0 \int_{K_{R_n}} \chi_{\{(u-k_n)_+>0\}} \, dx \, dt.$$

Recall that $a_0 = \left(\frac{\omega}{2^\lambda}\right)^{2-p}$ and divide by a_0 in the previous inequality to get

$$\sup_{-\frac{a_0}{2} R_n^p < t < 0} \int_{K_{R_n} \times \{t\}} (u - k_n)_+^p \zeta_n^p \, dx + \frac{1}{a_0} \int_{-\frac{a_0}{2} R_n^p}^0 \int_{K_{R_n}} |\nabla(u - k_n)_+ \zeta_n|^p \, dx \, dt$$

$$\leq \frac{C 2^{pn}}{R^p} \left(\frac{\omega}{2^\lambda}\right)^p \frac{1}{a_0} \int_{-\frac{a_0}{2} R_n^p}^0 \int_{K_{R_n}} \chi_{\{(u-k_n)_+>0\}} \, dx \, dt.$$

Next, perform a change in the time variable, putting

$$\bar{t} = \frac{t}{a_0/2}$$

and defining

$$\bar{u}(\cdot, \bar{t}) = u(\cdot, t) \qquad \text{and} \qquad \bar{\zeta}_n(\cdot, \bar{t}) = \zeta_n(\cdot, t),$$

to obtain the simplified inequality

$$\left\| (\bar{u} - k_n)_+ \bar{\zeta}_n \right\|_{V^p(Q(R_n^p, R_n))}^p \leq \frac{C 2^{pn}}{R^p} \left(\frac{\omega}{2^\lambda}\right)^p \int_{-R_n^p}^0 \int_{K_{R_n}} \chi_{\{(\bar{u}-k_n)_+>0\}} \, dx \, d\bar{t}.$$

Define, for each n,

$$A_n = \int_{-R_n^p}^0 \int_{K_{R_n}} \chi_{\{(\bar{u}-k_n)_+>0\}} \, dx \, d\bar{t}$$

and estimate, as before,

$$\frac{1}{2^{p(n+2)}} \left(\frac{\omega}{2^\lambda}\right)^p A_{n+1} = |k_{n+1} - k_n|^p A_{n+1}$$

$$\leq \left\| (\bar{u} - k_n)_+ \right\|_{p, Q(R_{n+1}^p, R_{n+1})}^p$$

$$\leq \left\| (\bar{u} - k_n)_+ \bar{\zeta}_n \right\|_{p, Q(R_n^p, R_n)}^p$$

$$\leq C \left\| (\bar{u} - k_n)_+ \bar{\zeta}_n \right\|_{V^p(Q(R_n^p, R_n))}^p A_n^{\frac{p}{d+p}}$$

$$\leq \frac{C 2^{pn}}{R^p} \left(\frac{\omega}{2^\lambda}\right)^p A_n^{1+\frac{p}{d+p}}. \tag{4.11}$$

Define the numbers

$$X_n = \frac{A_n}{|Q(R_n^p, R_n)|},$$

divide (4.11) by $|Q(R_{n+1}^p, R_{n+1})|$, and obtain the recursive relation

$$X_{n+1} \leq C\, 4^{pn}\, X_n^{1+\frac{p}{d+p}}.$$

By Lemma 2.9 on fast geometric convergence, if

$$X_0 \leq C^{-\frac{d+p}{p}}\, 4^{-\frac{(d+p)^2}{p}} =: \nu_* \tag{4.12}$$

then

$$X_n \longrightarrow 0. \tag{4.13}$$

Thus if (4.12) holds, then

$$\left| \left\{ (x,t) \in Q\left(\frac{a_0}{2}\left(\frac{R}{2}\right)^p, \frac{R}{2}\right) : u(x,t) > \mu^+ - \frac{\omega}{2^{\lambda+1}} \right\} \right| = 0$$

and the result follows. We are then left to prove (4.12).

4.3 Defining the Geometry

We now reach the crucial moment of fixing λ and, consequently, determining the length of the cylinder $Q(a_0 R^p, R)$ (recall the definition of a_0). The statement of Proposition 4.3 has then a double scope: we determine a level

$$\mu^+ - \frac{\omega}{2^{\lambda+1}}$$

and a cylinder (fixing λ and consequently a_0) such that the conclusion holds in that particular cylinder.

Proof of Proposition 4.3(continued). To simplify the symbolism introduce the sets

$$B_\sigma(t) = \left\{ x \in K_R : u(x,t) > \mu^+ - \frac{\omega}{2^\sigma} \right\}$$

and

$$B_\sigma = \left\{ (x,t) \in Q\left(\frac{a_0}{2} R^p, R\right) : u(x,t) > \mu^+ - \frac{\omega}{2^\sigma} \right\}.$$

With this notation, (4.12) reads

$$|B_\lambda| \leq \nu_* \left| Q\left(\frac{a_0}{2} R^p, R\right) \right|.$$

We will use the information contained in Corollary 4.2 to show that this holds, *i.e.*, that the subset of the cylinder $Q\left(\frac{a_0}{2} R^p, R\right)$ where u is close to its supremum μ^+ can be made arbitrarily small. Consider the local energy inequalities (2.6) for the functions $(u - k)_+$ in the cylinders $Q(a_0 R^p, 2R)$, with

$$k = \mu^+ - \frac{\omega}{2^s},$$

where s will be chosen later satisfying $s_2 \leq s \leq \lambda$ (recall that s_2 was chosen in Lemma 4.1). Take a piecewise smooth cutoff function $0 \leq \zeta \leq 1$, defined in $Q(a_0 R^p, 2R)$, and such that

$$\zeta = 1 \quad \text{in } Q\left(\frac{a_0}{2} R^p, R\right); \qquad \zeta = 0 \quad \text{on } \partial_p Q(a_0 R^p, 2R);$$

$$|\nabla \zeta| \leq \frac{1}{R}; \qquad\qquad 0 \leq \zeta_t \leq \frac{2}{a_0 R^p}.$$

Neglecting the first term on the left hand side of the estimates, we obtain for the indicated choices,

$$\int_{-\frac{a_0}{2} R^p}^0 \int_{K_R} |\nabla(u - k)_+|^p \, dx \, dt \leq \frac{C}{R^p} \int_{-a_0 R^p}^0 \int_{K_{2R}} (u - k)_+^p \, dx \, dt$$
$$+ \frac{C}{a_0 R^p} \int_{-a_0 R^p}^0 \int_{K_{2R}} (u - k)_+^2 \, dx \, dt.$$

We estimate the two terms on the right hand side of this inequality as follows:

$$\frac{C}{R^p} \int_{-a_0 R^p}^0 \int_{K_{2R}} (u - k)_+^p \, dx \, dt \leq \frac{C}{R^p} \left(\frac{\omega}{2^s}\right)^p \left| Q\left(\frac{a_0}{2} R^p, R\right) \right|$$

and, recalling the definition of a_0,

$$\frac{C}{a_0 R^p} \int_{-a_0 R^p}^0 \int_{K_{2R}} (u - k)_+^2 \, dx \, dt \leq \frac{C}{R^p} \left(\frac{\omega}{2^\lambda}\right)^{p-2} \left(\frac{\omega}{2^s}\right)^2 \left| Q\left(\frac{a_0}{2} R^p, R\right) \right|$$
$$\leq \frac{C}{R^p} \left(\frac{\omega}{2^s}\right)^p \left| Q\left(\frac{a_0}{2} R^p, R\right) \right|,$$

since $s \leq \lambda$. Gathering results, we reach

$$\iint_{B_s} |\nabla u|^p \, dx \, dt \leq \frac{C}{R^p} \left(\frac{\omega}{2^s}\right)^p \left| Q\left(\frac{a_0}{2} R^p, R\right) \right|. \tag{4.14}$$

We next apply Lemma 2.7 to the function $u(\cdot, t)$, for all $-\frac{a_0}{2} R^p \leq t \leq 0$, and with

$$k_1 = \mu^+ - \frac{\omega}{2^s}, \qquad k_2 = \mu^+ - \frac{\omega}{2^{s+1}}, \qquad k_2 - k_1 = \frac{\omega}{2^{s+1}}.$$

Observing that, owing to Corollary 4.2,

$$\left| \left\{ x \in K_R : u(x, t) \leq \mu^+ - \frac{\omega}{2^s} \right\} \right| = |K_R| - |B_s(t)| \geq \left(\frac{\nu_0}{2}\right)^2 |K_R|,$$

we obtain

$$\frac{\omega}{2^{s+1}}\,|B_{s+1}(t)| \le \frac{4C\,R^{d+1}}{\nu_0^2\,|K_R|}\int_{B_s(t)\setminus B_{s+1}(t)}|\nabla u|\,dx,$$

for $t \in \left(-\frac{a_0}{2}R^p,0\right)$. Integrating over this interval, we conclude that

$$\frac{\omega}{2^{s+1}}\,|B_{s+1}| \le \frac{CR}{\nu_0^2}\int\!\!\int_{B_s\setminus B_{s+1}}|\nabla u|\,dx\,dt$$

$$\le \frac{CR}{\nu_0^2}\left(\int\!\!\int_{B_s}|\nabla u|^p\,dx\,dt\right)^{\frac{1}{p}}|B_s\setminus B_{s+1}|^{\frac{p-1}{p}}$$

$$\le \frac{C}{\nu_0^2}\left(\frac{\omega}{2^s}\right)\left|Q\left(\frac{a_0}{2}R^p,R\right)\right|^{\frac{1}{p}}|B_s\setminus B_{s+1}|^{\frac{p-1}{p}},$$

using also (4.14). Simplifying and taking the $\frac{p}{p-1}$ power, we obtain

$$|B_{s+1}|^{\frac{p}{p-1}} \le C\,(\nu_0)^{-\frac{2p}{p-1}}\left|Q\left(\frac{a_0}{2}R^p,R\right)\right|^{\frac{1}{p-1}}|B_s\setminus B_{s+1}|.$$

Since these inequalities are valid for $s_2 \le s \le \lambda$, we add them for

$$s = s_2, s_2+1, s_2+2,\ldots,\lambda-1,$$

and since the sum on the right hand side is bounded above by $\left|Q\left(\frac{a_0}{2}R^p,R\right)\right|$, we obtain

$$(\lambda-s_2)\,|B_\lambda|^{\frac{p}{p-1}} \le C\,(\nu_0)^{-\frac{2p}{p-1}}\left|Q\left(\frac{a_0}{2}R^p,R\right)\right|^{\frac{p}{p-1}},$$

that is,

$$|B_\lambda| \le \frac{C}{(\lambda-s_2)^{\frac{p-1}{p}}}\,(\nu_0)^{-2}\left|Q\left(\frac{a_0}{2}R^p,R\right)\right|.$$

We obtain (4.12) if λ is chosen so large that

$$\frac{C}{\nu_0^2\,(\lambda-s_2)^{\frac{p-1}{p}}} \le \nu_*.$$

We finally make the choice

$$\lambda = \max\left\{s_2 + \left(\frac{C}{\nu_0^2\nu_*}\right)^{\frac{p}{p-1}},\, 1+\frac{1}{p-2}\right\} \tag{4.15}$$

(recall that s_2 is given through (4.7), ν_0 is given by (3.13), and ν_* is given by (4.12)) thus concluding the proof of the proposition. $\qquad\square$

Remark 4.4. Observe that the choice of λ was made so that (4.8) holds, and a larger λ is admissible.

Corollary 4.5. *Assume (4.1) and let (3.3) be in force. There exists a constant* $\sigma_1 \in (0,1)$, *depending only on the data, such that*

$$\underset{Q\left(\frac{a_0}{2}\left(\frac{R}{2}\right)^p, \frac{R}{2}\right)}{\text{ess osc}} \ u \ \leq \ \sigma_1 \, \omega.$$

Proof. It is similar to the proof of Corollary 3.5 for the choice

$$\sigma_1 = \left(1 - \frac{1}{2^{\lambda+1}}\right).$$

$\square$

4.4 The Recursive Argument

We finally prove the Hölder continuity of weak solutions through an iterative scheme designed from all previous results.

An immediate consequence of Corollaries 3.5 and 4.5 is the following.

Proposition 4.6. *There exists a constant* $\sigma \in (0,1)$, *that depends only on the data, such that*

$$\underset{Q\left(d\left(\frac{R}{8}\right)^p, \frac{R}{8}\right)}{\text{ess osc}} \ u \ \leq \ \sigma \, \omega.$$

Proof. Assume (3.3) is in force. Then, by Corollaries 3.5 and 4.5,

$$\underset{Q\left(d\left(\frac{R}{8}\right)^p, \frac{R}{8}\right)}{\text{ess osc}} \ u \ \leq \ \sigma \, \omega \,, \qquad \text{where} \quad \sigma = \max\{\sigma_0, \sigma_1\}, \qquad (4.16)$$

since

$$d\left(\frac{R}{8}\right)^p \leq \frac{a_0}{2}\left(\frac{R}{2}\right)^p.$$

$\square$

Proposition 4.7. *There exists a positive constant* C, *depending only on the data, such that, defining the sequences*

$$R_n = C^{-n}R \qquad\qquad \text{and} \qquad\qquad \omega_n = \sigma^n \omega,$$

for $n = 0, 1, 2, \ldots$, *where* $\sigma \in (0,1)$ *is given by (4.16), and constructing the family of cylinders*

$$Q_n = Q(a_n R_n^p, R_n) \,, \qquad \text{with} \qquad a_n = \left(\frac{\omega_n}{2^\lambda}\right)^{2-p},$$

where $\lambda > 1$ *is given by (4.15), we have*

$$Q_{n+1} \subset Q_n \qquad \text{and} \qquad \underset{Q_n}{\text{ess osc}} \ u \leq \omega_n, \qquad (4.17)$$

for all $n = 0, 1, 2, \ldots$

Proof. Recall the definition of $a_0 = \left(\frac{\omega}{2^\lambda}\right)^{2-p}$ and the construction of the initial cylinder so that the starting relation

$$\operatorname*{ess\,osc}_{Q_0} u \leq \omega \tag{4.18}$$

holds. We find

$$\begin{aligned}
d\left(\frac{R}{8}\right)^p &= \left(\frac{\omega}{2}\right)^{2-p}\frac{R^p}{8^p} \\
&= \left(\frac{\omega}{2}\right)^{2-p}\left(\frac{2^\lambda}{\omega_1}\right)^{2-p}\left(\frac{\omega_1}{2^\lambda}\right)^{2-p}\frac{R^p}{8^p} \\
&= \left(\frac{\omega}{\omega_1}\right)^{2-p}\left(\frac{2^\lambda}{2}\right)^{2-p}\left(\frac{\omega_1}{2^\lambda}\right)^{2-p}\frac{R^p}{8^p} \\
&= \sigma^{p-2}\,2^{(\lambda-1)(2-p)-3p}\,a_1\,R^p \\
&= a_1 R_1^p\,,
\end{aligned}$$

where $R_1 = C^{-1}R$, provided C is chosen from

$$C = \sigma^{\frac{2-p}{p}}\,2^{\frac{(\lambda-1)(p-2)}{p}+3} > 8\,.$$

From Proposition 4.6, we conclude

$$\operatorname*{ess\,osc}_{Q_1} u \leq \operatorname*{ess\,osc}_{Q\left(d\left(\frac{R}{8}\right)^p,\frac{R}{8}\right)} u \leq \sigma\,\omega = \omega_1,$$

which puts us back to the setting of (4.18). The entire process can now be repeated inductively starting from Q_1. □

Remark 4.8. The proof of Proposition 4.7 shows that it would have been sufficient to work with a number ω and a cylinder $Q(a_0 R^p, R)$ linked by (3.4). This relation is in general not verifiable *a priori* for a given cylinder, since its dimensions would have to be intrinsically defined in terms of the essential oscillation of u within it. The role of having introduced the cylinder $Q(R^2, R)$ and having assumed (3.3) is that (3.4) holds true for the *constructed* box $Q(a_0 R^p, R)$. It is part of the proof of proposition 4.7 to show that, at each step, the cylinders Q_n and the essential oscillation of u within them satisfy the intrinsic geometry dictated by (3.4).

Lemma 4.9. *There exist constants $\gamma > 1$ and $\alpha \in (0,1)$, that can be determined a priori in terms of the data, such that for all the cylinders*

$$Q(a_0\rho^p, \rho)\,, \qquad with \quad 0 < \rho \leq R\,,$$
$$\operatorname*{ess\,osc}_{Q(a_0\rho^p,\rho)} u \leq \gamma\,\omega\left(\frac{\rho}{R}\right)^\alpha.$$

Proof. Let $0 < \rho \leq R$ be fixed. There exists a non-negative integer n such that

$$C^{-(n+1)}R \leq \rho < C^{-n}R$$

so, putting $\alpha = -\dfrac{\ln\sigma}{\ln C}$, we deduce

$$C^{-(n+1)} \leq \frac{\rho}{R} \iff \sigma^{\frac{n+1}{\alpha}} \leq \frac{\rho}{R} \iff \sigma^{n+1} \leq \left(\frac{\rho}{R}\right)^{\alpha}.$$

Thus

$$\omega_n = \sigma^n \omega \leq \gamma\, \omega \left(\frac{\rho}{R}\right)^{\alpha}, \qquad \text{with} \quad \gamma = \sigma^{-1}.$$

To conclude the proof, observe that the cylinder $Q(a_0\rho^p, \rho)$ is contained in the cylinder $Q_n = Q(a_n R_n^p, R_n)$, since

$$\omega_n \leq \omega \qquad \text{and} \qquad \rho < C^{-n}R = R_n.$$

$\square$

Let $\Gamma = \partial_p \Omega_T$ be the parabolic boundary of Ω_T and u be a bounded local weak solution of (2.1) in Ω_T, with $M = \|u\|_{\infty,\Omega_T}$. Introduce the degenerate intrinsic parabolic p-distance from a compact set $K \subset \Omega_T$ to Γ, by

$$p - \text{dist}(K; \Gamma) := \inf_{\substack{(x,t)\in K \\ (y,s)\in \Gamma}} \left(|x - y| + M^{\frac{p-2}{p}}|t - s|^{\frac{1}{p}}\right).$$

Theorem 4.10. *Let u be a bounded local weak solution of (2.1) in Ω_T and $M = \|u\|_{\infty,\Omega_T}$. Then u is locally Hölder continuous in Ω_T, i.e., there exist constants $\gamma > 1$ and $\alpha \in (0,1)$, depending only on the data, such that, for every compact subset K of Ω_T,*

$$|u(x_1,t_1) - u(x_2,t_2)| \leq \gamma M \left(\frac{|x_1 - x_2| + M^{\frac{p-2}{p}}|t_1 - t_2|^{\frac{1}{p}}}{p - \text{dist}(K; \Gamma)}\right)^{\alpha},$$

for every pair of points $(x_i, t_i) \in K$, $i = 1, 2$.

Proof. Fix $(x_i, t_i) \in K$, $i = 1, 2$, such that $t_2 > t_1$ and construct the cylinder

$$(x_2, t_2) + Q\left(M^{2-p}R^p, R\right).$$

It is contained in Ω_T if we choose

$$R \leq \inf_{\substack{x\in K \\ y\in\partial\Omega}} |x - y| \qquad \text{and} \qquad M^{\frac{2-p}{p}}R \leq \inf_{t\in K} t^{\frac{1}{p}}.$$

Thus, in particular, we may choose $2R = p - \text{dist}(K; \Gamma)$. To prove the Hölder continuity in the t–variable assume first that

$$t_2 - t_1 < M^{2-p}R^p.$$

Then, there exists $\rho \in (0, R)$ such that $t_2 - t_1 = M^{2-p}\rho^p$, i.e.,

$$\rho = M^{\frac{p-2}{p}}|t_2 - t_1|^{\frac{1}{p}}.$$

The oscillation inequality of Lemma 4.9, applied in the cylinder

$$(x_2, t_2) + Q(a_0\rho^p, \rho)$$

implies

$$|u(x_2, t_2) - u(x_2, t_1)| \leq \gamma M \left(\frac{M^{\frac{p-2}{p}}|t_2 - t_1|^{\frac{1}{p}}}{p - \mathrm{dist}(K; \Gamma)} \right)^\alpha.$$

If $(t_2 - t_1) \geq M^{2-p}R^p$, we have

$$|u(x_2, t_2) - u(x_2, t_1)| \leq 2M \leq 4M \left(\frac{M^{\frac{p-2}{p}}|t_2 - t_1|^{\frac{1}{p}}}{p - \mathrm{dist}(K; \Gamma)} \right).$$

The Hölder continuity in the space variables is proved analogously.   □

Remark 4.11. The theory includes statements of regularity up to the parabolic boundary of Ω_T (cf. [14] and [52]).

4.5 Generalizations

The analysis of the singular case $1 < p < 2$ is somehow more involved, but several of the previous techniques apply (see [14] and [21]).

As indicated earlier, the Hölder continuity of u is solely a consequence of the energy inequalities (2.6) and the logarithmic inequalities (2.8). For this reason, the techniques just presented are rather flexible and adjust to a variety of singular and degenerate parabolic partial differential equations.

A possible generalization is to equations with the full p–Laplacian type quasilinear structure

$$u_t - \mathrm{div}\, \mathbf{a}(x, t, u, \nabla u) = b(x, t, u, \nabla u) \qquad \text{in } \mathcal{D}'(\Omega_T), \tag{4.19}$$

where $\mathbf{a} : \Omega_T \times \mathbb{R}^{d+1} \to \mathbb{R}^d$ and $b : \Omega_T \times \mathbb{R}^{d+1} \to \mathbb{R}$ are measurable and satisfy the structure assumptions

(A_1) $\mathbf{a}(x, t, u, \nabla u) \cdot \nabla u \geq C_0|\nabla u|^p - \varphi_0(x, t);$
(A_2) $|\mathbf{a}(x, t, u, \nabla u)| \leq C_1|\nabla u|^{p-1} + \varphi_1(x, t);$
(A_3) $|b(x, t, u, \nabla u)| \leq C_2|\nabla u|^p + \varphi_2(x, t),$

for a.e. $(x, t) \in \Omega_T$, with $p > 1$. The C_i, $i = 0, 1, 2$, are given positive constants and the φ_i, $i = 0, 1, 2$, are given non-negative functions, defined in Ω_T and subject to the integrability conditions

$$\varphi_0, \quad \varphi_1^{\frac{p}{p-1}}, \quad \varphi_2 \in L^{q,r}(\Omega_T)$$

with $q, r \geq 1$ satisfying, for $1 < p \leq d$,

$$\frac{1}{r} + \frac{d}{pq} \in (0,1).$$

See [14] for the details.

Another family of equations to which the theory applies are degenerate or singular equations of porous medium type that can be cast in the form (4.19), for the structure assumptions

(B$_1$) $\mathbf{a}(x,t,u,\nabla u) \cdot \nabla u \geq C_0 |u|^{m-1} |\nabla u|^2 - \varphi_0(x,t),$ $m > 0$;
(B$_2$) $|\mathbf{a}(x,t,u,\nabla u)| \leq C_1 |u|^{m-1} |\nabla u| + \varphi_1(x,t);$
(B$_3$) $|b(x,t,u,\nabla u)| \leq C_2 |\nabla|u|^m|^2 + \varphi_2(x,t),$

and the functions φ_i, $i = 0,1,2$, satisfy the same conditions as before with $p = 2$. We require

$$u \in L^\infty_{\text{loc}}\left(0,T; L^2_{\text{loc}}(\Omega)\right) \qquad \text{and} \qquad |u|^m \in L^2_{\text{loc}}\left(0,T; W^{1,2}_{\text{loc}}(\Omega)\right).$$

There is a wide literature concerning this problem and we refer the reader to [3, 55] and the references therein. See also Chapter 6.

Further generalizations can be obtained by replacing s^{m-1}, $s > 0$, with a function that blows up like a power when $s \searrow 0$ and is regular otherwise. To be specific, consider doubly degenerate equations of the form (4.19) with structure assumptions

(C$_1$) $\mathbf{a}(x,t,u,\nabla u) \cdot \nabla u \geq C_0 \Phi(|u|)|\nabla u|^p - \varphi_0(x,t);$
(C$_2$) $|\mathbf{a}(x,t,u,\nabla u)| \leq C_1 \Phi(|u|)|\nabla u|^{p-1} + \Phi^{\frac{1}{p}}(u)\varphi_1(x,t);$
(C$_3$) $|b(x,t,u,\nabla u)| \leq C_2 \Phi(|u|)|\nabla u|^p + \varphi_2(x,t).$

Here φ_i, $i = 0,1,2$, satisfy the same conditions as before and the function $\Phi(\cdot)$ is degenerate near the origin in the sense that

$$\exists \sigma > 0 \; : \; \gamma_1 s^{\beta_1} \leq \Phi(s) \leq \gamma_2 s^{\beta_2}, \qquad \forall s \in (0,\sigma),$$

for given constants $0 < \gamma_1 \leq \gamma_2$ and $0 \leq \beta_2 \leq \beta_1$. For $s > \sigma$, i.e., away from zero, it is assumed that Φ is bounded above and below by given positive constants. We require that

$$u \in C_{\text{loc}}\left(0,T; L^2_{\text{loc}}(\Omega)\right) \qquad \text{and} \qquad \Phi^{\frac{1}{p-1}}(u)|\nabla u| \in L^p_{\text{loc}}(\Omega_T)$$

and, denoting with $F(\cdot)$ the primitive of $\Phi^{\frac{1}{p-1}}(\cdot)$, that

$$F(u) \in L^p_{\text{loc}}\left(0,T; W^{1,p}_{\text{loc}}(\Omega)\right),$$

which allows for an interpretation of the equation in the weak sense. One recognizes that if $\Phi(s) \equiv 1$ the equation is of p-Laplacian type and if $\Phi(s) = s^{m-1}$ and $p = 2$ the equation is of porous medium type. The Hölder continuity of solutions was obtained independently in [46] and [31].

Part II

Some Applications

5

Immiscible Fluids and Chemotaxis

This second part is devoted to a series of three applications of the method of intrinsic scaling to relevant models arising from flows in porous media, chemotaxis and phase transitions.

We start with the flow of two immiscible fluids through a porous medium, proving the Hölder continuity of the saturations, which satisfy a PDE with a two-sided degeneracy. The same type of structure arises in a model for the chemotactic movement of cells under a volume-filling effect and the extension to this case, which basically consists in dealing appropriately with an extra lower order term, is also included.

5.1 The Flow of Two Immiscible Fluids through a Porous Medium

The flow of two immiscible fluids through a porous medium can be modeled through a parabolic equation with two degeneracies, namely

$$v_t - \operatorname{div}(\alpha(v)\nabla v) = 0, \tag{5.1}$$

where $v \in [0, 1]$ and $\alpha(v)$ degenerates for $v = 0$ and $v = 1$.

Following the important pioneering work in [35], that deals with the case of a strictly parabolic equation, a mathematical analysis of this model, where v represents the saturation of one of the fluids, was developed in [1]. The existence of a weak solution for the problem was established, together with the continuity of the saturation v under the assumption that α degenerates at most at one side, although no restrictions were put on the nature of the degeneracy. The result was extended to a setting where two degeneracies for α are allowed, provided some information on the nature of one of the degeneracies is assumed. First, in the same paper, the decay of α at one side was taken at most logarithmic and later, in [13], α was allowed to behave like a power. In [53] it was shown that v is locally Hölder continuous if α decays like a power at *both* degeneracies. We stress that the novelty lies in the Hölder character of the solution since the continuity is well known. In fact, equation (5.1) can be written in the form

$$[\beta(w)]_t - \Delta w = 0 \qquad \text{with} \qquad w = \int_0^v \alpha(s)\, ds\;; \quad v = \beta(w),$$

where β is an increasing function, for which the results of [22] apply. Some further physical motivation for the study of this type of equations, arising in polymer chemistry and combustion models, may be found in [30], where a numerical scheme is used to compute approximate solutions and interface curves for the Cauchy problem associated to (5.1) in a one dimensional case.

From the mathematical point of view, the equation is interesting in that it exhibits a strong smoothing effect that prevents the possible development of singularities due to the degeneracies. The ultimate goal would be to obtain the continuity for a general quasi-linear equation, irrespective of the nature of the two degeneracies. This is also relevant in physical terms since the available experimental information about α is only of a qualitative nature, which makes all assumptions on the degeneracies quite restrictive (see [15]).

We will consider here local weak solutions of equation (5.1), assuming that they exist; for the existence theory see [1].

Definition 5.1. *A local weak solution of (5.1) is a measurable function $v(x,t)$ defined in Ω_T and such that*

(i) $v \in [0,1]$ and $v \in C\left(0,T;L^2(\Omega)\right)$;

(ii) $\displaystyle\int_0^v \alpha(s)\, ds \ \in\ L^2\left(0,T;H^1_{\text{loc}}(\Omega)\right)$;

(iii) *for every subset $K \subset \Omega$ and for every subinterval $[t_1,t_2] \subset (0,T)$,*

$$\int_K v\varphi\, dx\, \Big|_{t_1}^{t_2} + \int_{t_1}^{t_2}\int_K \{-v\,\varphi_t + \alpha(v)\,\nabla v \cdot \nabla\varphi\}\, dx\, dt = 0,$$

for all $\varphi \in H^1_{\text{loc}}\left(0,T;L^2(K)\right) \cap L^2_{\text{loc}}\left(0,T;H^1_0(K)\right)$.

We can write (iii) in an equivalent way that is technically more convenient and involves the discrete time derivative. Recalling definition (2.3) of the Steklov average of a function, the equivalent formulation reads

(iii)′ *for every compact $K \subset \Omega$ and for every $0 < t < T - h$,*

$$\int_{K\times\{t\}} \{(v_h)_t\, \varphi + (\alpha(v)\nabla v)_h \cdot \nabla\varphi\}\, dx = 0, \tag{5.2}$$

for all $\varphi \in H^1_0(K)$.

The regularity result will be obtained under the following assumptions on the diffusion coefficient α:

(A1) α is a continuous function and $\alpha(v) > 0$, for $v \in (0, 1)$.

(A2) $\exists\, \delta_0 \in (0, \frac{1}{2})$ such that, for constants $1 < C_0 < C_1$,

$$C_0\, \phi(v) \leq \alpha(v) \leq C_1\, \phi(v)\,, \quad \forall v \in [0, \delta_0]$$

and

$$C_0\, \psi(1 - v) \leq \alpha(v) \leq C_1\, \psi(1 - v)\,, \quad \forall v \in [1 - \delta_0, 1].$$

(A3) $\psi(s) = s^{p_1}$; $\phi(s) = s^{p_2}$; $p_1 > p_2 > 0$.

Although we are working with powers, the proof carries through with power-like ϕ and ψ, $i.e.$, functions that satisfy a condition of the type

$$\phi(v) \geq C\, v\, \phi'(v).$$

Our main result is stated next.

Theorem 5.2. *Under assumptions (A1)-(A3), any local weak solution of (5.1) is locally Hölder continuous.*

The proof of the local Hölder continuity uses the method of intrinsic scaling described in Part I. We will apply it here in a new setting since the nature of the degeneracies is quite different from the case of the p–Laplacian. For that reason, this chapter was kept essentially self-contained and can be read independently of most of Part I.

5.2 Rescaled Cylinders

Consider a point $(x_0, t_0) \in \Omega_T$ and, by translation and to simplify, assume $(x_0, t_0) = (0, 0)$. Consider a small positive number $\epsilon > 0$ and $R > 0$ such that

$$Q((2R)^{2-\epsilon}, 2R) \subset \Omega_T$$

and define

$$\mu^- := \operatorname*{ess\,inf}_{Q((2R)^{2-\epsilon}, 2R)} v \; ; \quad \mu^+ := \operatorname*{ess\,sup}_{Q((2R)^{2-\epsilon}, 2R)} v \; ;$$

$$\omega := \operatorname*{ess\,osc}_{Q((2R)^{2-\epsilon}, 2R)} v = \mu^+ - \mu^- \; .$$

Construct the cylinder

$$Q(\theta R^2, R), \quad \text{with} \quad \theta^{-1} = \phi\left(\frac{\omega}{2^m}\right),$$

where the number m will be chosen large, later in the proof, independently of ω. We assume $\mu^+ = 1$ and $\mu^- = 0$, since the other possibilities are clearly more favorable.

We may assume that $Q(\theta R^2, R) \subset Q((2R)^{2-\epsilon}, 2R)$, which means that

$$-\theta R^2 \geq -(2R)^{2-\epsilon} \quad \Longleftrightarrow \quad \theta^{-1} \geq 2^{\epsilon-2} R^\epsilon. \tag{5.3}$$

If this does not hold, then we have

$$\phi\left(\frac{\omega}{2^m}\right) < CR^\epsilon,$$

and then the oscillation would go to zero with R and there would be nothing to prove. We then have the relation

$$\operatorname*{ess\,osc}_{Q(\theta R^2, R)} v \leq \omega \tag{5.4}$$

which will be the starting point of the iteration process that leads to our main result. Note that we had to consider the cylinder $Q((2R)^{2-\epsilon}, 2R)$ and assume (5.3), so that (5.4) would hold for the rescaled cylinder $Q(\theta R^2, R)$. This is in general not true for a given cylinder since its dimensions would have to be intrinsically defined in terms of the essential oscillation of the function within it. Observe also that when the oscillation ω is small, and for m very large, then the cylinder $Q(\theta R^2, R)$ is very long in the t direction. It is this feature that will allow us to accommodate the two degeneracies in the problem. We will also assume, without loss of generality, that $\omega < \delta_0$, where δ_0 was introduced in (A2).

We now consider subcylinders of $Q(\theta R^2, R)$ of the form

$$Q_R^{t^*} := (0, t^*) + Q\left(\frac{R^2}{\psi(\frac{\omega}{4})}, R\right), \quad \text{with} \quad t^* < 0.$$

They are contained in $Q(\theta R^2, R)$ if $\theta R^2 \geq -t^* + \frac{R^2}{\psi(\frac{\omega}{4})}$, which holds if

$$\phi\left(\frac{\omega}{2^m}\right) \leq \psi(\frac{\omega}{4})$$

and t^* is chosen such that

$$t^* \in \left(\frac{R^2}{\psi(\frac{\omega}{4})} - \frac{R^2}{\phi(\frac{\omega}{2^m})}, 0\right). \tag{5.5}$$

We will assume further, and for technical reasons, that

$$\phi\left(\frac{\omega}{2^m}\right) \leq \frac{1}{2}\psi\left(\frac{\omega}{4}\right) \tag{5.6}$$

emphasizing that it is the choice of a big cylinder $Q(\theta R^2, R)$, by choosing m very large, that makes (5.6) possible.

The proof of the Hölder continuity follows from the analysis of an alternative. We first concentrate on the degeneracy at 1 and later consider the behavior of v near 0. The common bottom line will be that going down to a smaller cylinder the oscillation decreases by a small factor that we can exhibit and that does not depend on the oscillation.

5.3 Focusing on One Degeneracy

For a constant $\nu_0 \in (0,1)$, that will be determined depending only on the data, we will assume in this section that there is a cylinder of the type $Q_R^{t^*}$ for which

$$\left|\left\{(x,t) \in Q_R^{t^*} : v(x,t) > 1 - \frac{\omega}{2}\right\}\right| \leq \nu_0 \left|Q_R^{t^*}\right| \tag{5.7}$$

leaving for the next section the analysis of the complementary case. We start by showing that if (5.7) holds then v is away from the degeneracy at 1 in a smaller cylinder of the same type. The next lemma specifies what this means.

Lemma 5.3. *There exists a constant $\nu_0 \in (0,1)$, depending only on the data, such that if (5.7) holds then*

$$v(x,t) < 1 - \frac{\omega}{4}, \quad a.e. \ in \ Q_{\frac{R}{2}}^{t^*}.$$

Proof. Let $v_\omega := \min\{v, 1 - \frac{\omega}{4}\}$. Take the cylinder for which (5.7) holds, define

$$R_n = \frac{R}{2} + \frac{R}{2^{n+1}}, \quad n = 0, 1, \ldots,$$

and construct the family of nested and shrinking cylinders

$$Q_{R_n}^{t^*} = K_{R_n} \times \left(t^* - \frac{R_n^2}{\psi(\frac{\omega}{4})}, t^*\right).$$

Consider piecewise smooth cutoff functions $0 \leq \xi_n \leq 1$, defined in these cylinders, and satisfying the following set of assumptions

$$\xi_n = 1 \ \text{in} \ Q_{R_{n+1}}^{t^*}; \qquad\qquad\qquad \xi_n = 0 \ \text{on} \ \partial_p Q_{R_n}^{t^*};$$

$$|\nabla \xi_n| \leq \frac{2^{n+1}}{R}; \tag{5.8}$$

$$|\Delta \xi_n| \leq \frac{2^{2(n+1)}}{R^2}; \qquad\qquad 0 \leq (\xi_n)_t \leq 2^{2(n+1)}\frac{\psi(\frac{\omega}{4})}{R^2}.$$

Let

$$k_n = 1 - \frac{\omega}{4} - \frac{\omega}{2^{n+2}}, \quad n = 0, 1, \dots$$

choose as test function in (5.2) $\varphi = [(v_\omega)_h - k_n]_+ \xi_n^2$ and integrate in time over $(t^* - \frac{R_n^2}{\psi(\frac{\omega}{4})}, t)$ for $t \in (t^* - \frac{R_n^2}{\psi(\frac{\omega}{4})}, t^*)$. Putting

$$\tau_n := t^* - \frac{R_n^2}{\psi(\frac{\omega}{4})}$$

the first term gives (for $K = K_{R_n}$ and omitting dx and dt throughout)

$$\int_{\tau_n}^t \int_{K_{R_n}} (v_h)_t \, [(v_\omega)_h - k_n]_+ \, \xi_n^2 = \frac{1}{2} \int_{\tau_n}^t \int_{K_{R_n}} \left([(v_\omega)_h - k_n]_+^2 \right)_t \xi_n^2$$

$$+ \left(1 - \frac{\omega}{4} - k_n \right) \int_{\tau_n}^t \int_{K_{R_n}} \left(\left(\left[v - (1 - \frac{\omega}{4}) \right]_+ \right)_h \right)_t \xi_n^2.$$

Next, integrate by parts and let $h \to 0$. Using the properties of the Steklov average, we get

$$\frac{1}{2} \int_{K_{R_n} \times \{t\}} (v_\omega - k_n)_+^2 \, \xi_n^2 - \frac{1}{2} \int_{K_{R_n} \times \{\tau_n\}} (v_\omega - k_n)_+^2 \, \xi_n^2$$

$$- \int_{\tau_n}^t \int_{K_{R_n}} (v_\omega - k_n)_+^2 \, \xi_n (\xi_n)_t$$

$$+ \left(1 - \frac{\omega}{4} - k_n \right) \left\{ \int_{K_{R_n} \times \{t\}} \left[v - \left(1 - \frac{\omega}{4} \right) \right]_+ \xi_n^2 \right.$$

$$- \int_{K_{R_n} \times \{\tau_n\}} \left[v - \left(1 - \frac{\omega}{4} \right) \right]_+ \xi_n^2$$

$$\left. - 2 \int_{\tau_n}^t \int_{K_{R_n}} \left[v - \left(1 - \frac{\omega}{4} \right) \right]_+ \xi_n (\xi_n)_t \right\}$$

Since the second and the fifth terms vanish, due to the fact that ξ_n was chosen such that it vanishes on the parabolic boundary of $Q_{R_n}^{t^*}$, and the fourth term is positive, we estimate from below, using the other assumptions on ξ_n, by

$$\frac{1}{2} \int_{K_{R_n} \times \{t\}} (v_\omega - k_n)_+^2 \, \xi_n^2 - \psi\left(\frac{\omega}{4} \right) \frac{2^{2(n+1)}}{R^2} \left(\frac{\omega}{4} \right)^2 \int_{\tau_n}^t \int_{K_{R_n}} \chi_{\{v_\omega \geq k_n\}}$$

$$- 2\psi\left(\frac{\omega}{4} \right) \frac{2^{2(n+1)}}{R^2} \left(\frac{\omega}{4} \right)^2 \int_{\tau_n}^t \int_{K_{R_n}} \chi_{\{v \geq 1 - \frac{\omega}{4}\}} = (*).$$

Observe that

$$(v_\omega - k_n)_+ \le \frac{\omega}{4}, \qquad 1 - \frac{\omega}{4} - k_n \le \frac{\omega}{4} \qquad \text{and} \qquad \left[v - \left(1 - \frac{\omega}{4}\right)\right]_+ \le \frac{\omega}{4}.$$

Finally, remarking that $v \ge 1 - \frac{\omega}{4} \Rightarrow v_\omega \ge k_n$, we obtain

$$(*) \ge \frac{1}{2} \int_{K_{R_n} \times \{t\}} (v_\omega - k_n)_+^2 \, \xi_n^2 - 3\psi\left(\frac{\omega}{4}\right) \frac{2^{2(n+1)}}{R^2} \left(\frac{\omega}{4}\right)^2 \int_{\tau_n}^t \int_{K_{R_n}} \chi_{\{v_\omega \ge k_n\}}.$$

Concerning the diffusion term, we first pass to the limit in h, obtaining

$$\int_{\tau_n}^t \int_{K_{R_n}} (\alpha(v)\nabla v)_h \cdot \nabla \left\{[(v_\omega)_h - k_n]_+ \, \xi_n^2\right\}$$

$$\longrightarrow \int_{\tau_n}^t \int_{K_{R_n}} \alpha(v)\nabla v \cdot \left\{\xi_n^2 \nabla(v_\omega - k_n)_+ + 2(v_\omega - k_n)_+ \xi_n \nabla \xi_n\right\}$$

$$= \int_{\tau_n}^t \int_{K_{R_n}} \alpha(v) \, |\xi_n \nabla(v_\omega - k_n)_+|^2 + 2 \int_{\tau_n}^t \int_{K_{R_n}} \alpha(v)(v_\omega - k_n)_+ \xi_n \, \nabla v \cdot \nabla \xi_n.$$

Next, we estimate the second term:

$$\left| 2 \int_{\tau_n}^t \int_{K_{R_n}} \alpha(v)(v_\omega - k_n)_+ \xi_n \, \nabla v \cdot \nabla \xi_n \right|$$

$$\le 2 \int_{\tau_n}^t \int_{K_{R_n}} |\alpha(v)| \, (v_\omega - k_n)_+ \, \xi_n \, |\nabla \xi_n| \, |\nabla(v_\omega - k_n)_+|$$

$$+ \left| 2\left(1 - \frac{\omega}{4} - k_n\right) \int_{\tau_n}^t \int_{K_{R_n}} \xi_n \, \nabla \left\{ \left(\int_{1-\frac{\omega}{4}}^v \alpha(s) \, ds\right)_+ \right\} \cdot \nabla \xi_n \right|$$

$$\le C_1 \psi\left(\frac{\omega}{2}\right) \left\{ \epsilon \int_{\tau_n}^t \int_{K_{R_n}} |\xi_n \nabla(v_\omega - k_n)_+|^2 + \frac{1}{\epsilon} \int_{\tau_n}^t \int_{K_{R_n}} (v_\omega - k_n)_+^2 \, |\nabla \xi_n|^2 \right\}$$

$$+ 2\left(\frac{\omega}{4}\right) \left| -\int_{\tau_n}^t \int_{K_{R_n}} \left(\int_{1-\frac{\omega}{4}}^v \alpha(s) \, ds\right)_+ (|\nabla \xi_n|^2 + \xi_n \Delta \xi_n) \right|$$

$$\le C_1 \epsilon \psi\left(\frac{\omega}{2}\right) \int_{\tau_n}^t \int_{K_{R_n}} |\xi_n \nabla(v_\omega - k_n)_+|^2$$

$$+ \frac{C_1 2^{2(n+1)} \psi(\frac{\omega}{2})}{\epsilon R^2} \left(\frac{\omega}{4}\right)^2 \int_{\tau_n}^t \int_{K_{R_n}} \chi_{\{v_\omega \ge k_n\}}$$

$$+ 2\left(\frac{\omega}{4}\right) 2 \frac{2^{2(n+1)}}{R^2} \psi\left(\frac{\omega}{4}\right) \left(\frac{\omega}{4}\right) \int_{\tau_n}^t \int_{K_{R_n}} \chi_{\{v_\omega \ge k_n\}},$$

since

$$\left(\int_{1-\frac{\omega}{4}}^{v} \alpha(s)\,ds\right)_+ \le \psi\left(1-(1-\frac{\omega}{4})\right)\left(1-(1-\frac{\omega}{4})\right).$$

Observing that $\nabla(v_\omega - k_n)_+$ is only nonzero in the set $\{k_n < v < 1 - \frac{\omega}{4}\}$, and that in this set

$$\alpha(v) \ge C_0\,\psi(1-v) \ge C_0\,\psi\left(1-(1-\frac{\omega}{4})\right) = C_0\,\psi\left(\frac{\omega}{4}\right),$$

we conclude, choosing

$$\epsilon = \frac{(C_0 - \frac{1}{2})\psi(\frac{\omega}{4})}{C_1\psi(\frac{\omega}{2})},$$

that

$$\int_{\tau_n}^{t}\int_{K_{R_n}} \alpha(v)\,\nabla v \cdot \nabla\left[(v_\omega - k_n)_+\xi_n^2\right] \ge \frac{1}{2}\psi\left(\frac{\omega}{4}\right)\int_{\tau_n}^{t}\int_{K_{R_n}} |\xi_n\nabla(v_\omega - k_n)_+|^2$$
$$-\left\{\frac{[C_1\psi(\frac{\omega}{2})]^2}{(C_0 - \frac{1}{2})\psi(\frac{\omega}{4})} + 4\psi\left(\frac{\omega}{4}\right)\right\}\frac{2^{2(n+1)}}{R^2}\left(\frac{\omega}{4}\right)^2\int_{\tau_n}^{t}\int_{K_{R_n}} \chi_{\{v_\omega \ge k_n\}}.$$

Now, putting both estimates together, we arrive at

$$\operatorname*{ess\,sup}_{\tau_n \le t \le t^*}\int_{K_{R_n}\times\{t\}} (v_\omega - k_n)_+^2\,\xi_n^2 + \psi\left(\frac{\omega}{4}\right)\int_{\tau_n}^{t^*}\int_{K_{R_n}} |\xi_n\nabla(v_\omega - k_n)_+|^2$$
$$\le 2\left\{\frac{[C_1\psi(\frac{\omega}{2})]^2}{(C_0 - \frac{1}{2})\psi(\frac{\omega}{4})} + 7\psi\left(\frac{\omega}{4}\right)\right\}\frac{2^{2(n+1)}}{R^2}\left(\frac{\omega}{4}\right)^2\int_{\tau_n}^{t^*}\int_{K_{R_n}} \chi_{\{v_\omega \ge k_n\}}.$$

Next we perform a change in the time variable, putting $\bar{t} = (t - t^*)\psi(\frac{\omega}{4})$ and define

$$\overline{v_\omega}(\cdot, \bar{t}) = v_\omega(\cdot, t) \qquad \text{and} \qquad \overline{\xi_n}(\cdot, \bar{t}) = \xi_n(\cdot, t),$$

obtaining the simplified inequality

$$\left\|(\overline{v_\omega} - k_n)_+\,\overline{\xi_n}\right\|_{V^2(Q(R_n^2, R_n))}^2$$
$$\le 2\left\{\frac{C_1^2}{(C_0 - \frac{1}{2})}\left[\frac{\psi(\frac{\omega}{2})}{\psi(\frac{\omega}{4})}\right]^2 + 7\right\}\frac{2^{2(n+1)}}{R^2}\left(\frac{\omega}{4}\right)^2\int_{-R_n^2}^{0}\int_{K_{R_n}} \chi_{\{\overline{v_\omega} \ge k_n\}}. \qquad (5.9)$$

Define, for each n,

$$A_n = \int_{-R_n^2}^{0}\int_{K_{R_n}} \chi_{\{\overline{v_\omega} \ge k_n\}}\,dx\,d\bar{t}$$

and observe that the following estimates hold

$$\frac{1}{2^{2(n+2)}} \left(\frac{\omega}{4}\right)^2 A_{n+1} = |k_{n+1} - k_n|^2 A_{n+1}$$

$$\leq \left\| (\overline{v_\omega} - k_n)_+ \right\|_{2, Q(R_{n+1}^2, R_{n+1})}^2$$

$$\leq \left\| (\overline{v_\omega} - k_n)_+ \overline{\xi_n} \right\|_{2, Q(R_n^2, R_n)}^2$$

$$\leq C \left\| (\overline{v_\omega} - k_n)_+ \overline{\xi_n} \right\|_{V^2(Q(R_n^2, R_n))}^2 A_n^{\frac{2}{d+2}}$$

$$\leq 2C \left\{ \frac{C_1^2}{(C_0 - \frac{1}{2})} \left[\frac{\psi(\frac{\omega}{2})}{\psi(\frac{\omega}{4})} \right]^2 + 7 \right\} \times \frac{2^{2(n+1)}}{R^2} \left(\frac{\omega}{4}\right)^2 A_n^{1+\frac{2}{d+2}}. \qquad (5.10)$$

In fact, the second inequality is obvious; the first one holds due to the fact that $k_n < k_{n+1}$; the third inequality is a consequence of Theorem 2.11, and the last one follows from (5.9). Next, define the numbers

$$X_n = \frac{A_n}{|Q(R_n^2, R_n)|},$$

divide (5.10) by $|Q(R_{n+1}^2, R_{n+1})|$ and obtain the recursive relation

$$X_{n+1} \leq \gamma \, 4^{2n} \, X_n^{1+\frac{2}{d+2}},$$

where

$$\gamma = C \left\{ \frac{C_1^2}{(C_0 - \frac{1}{2})} \left[\frac{\psi(\frac{\omega}{2})}{\psi(\frac{\omega}{4})} \right]^2 + 7 \right\}.$$

We can use Lemma 2.9 to conclude that if

$$X_0 \leq \gamma^{-\frac{d+2}{2}} 4^{-2(\frac{d+2}{2})^2} =: \nu_0 \qquad (5.11)$$

then

$$X_n \longrightarrow 0. \qquad (5.12)$$

But (5.11) is nothing but the assumption (5.7) of the lemma and the conclusion easily follows from (5.12). In fact, observe that

$$R_n \searrow \frac{R}{2} \qquad \text{and} \qquad k_n \nearrow 1 - \frac{\omega}{4},$$

and since (5.12) implies that $A_n \to 0$, we conclude that

$$\left| \left\{ (x, \overline{t}) \in Q\left(\left(\frac{R}{2}\right)^2, \frac{R}{2} \right) : \overline{v_\omega}(x, \overline{t}) \geq 1 - \frac{\omega}{4} \right\} \right|$$

$$= \left| \left\{ (x, t) \in Q_{\frac{R}{2}}^{t^*} : v(x, t) \geq 1 - \frac{\omega}{4} \right\} \right| = 0$$

and the lemma is proved.

Let us just show why ν_0 is independent of ω, since this is crucially related to the fact that ψ is a power. In fact, we have

$$\nu_0 = \gamma^{-\frac{d+2}{2}} 4^{-2(\frac{d+2}{2})^2} = 4^{-2(\frac{d+2}{2})^2} C \left\{ \frac{C_1^2}{(C_0 - \frac{1}{2})} \left[\frac{\psi(\frac{\omega}{2})}{\psi(\frac{\omega}{4})} \right]^2 + 7 \right\}^{-\frac{d+2}{2}},$$

and so we conclude showing that

$$\frac{\psi(\frac{\omega}{2})}{\psi(\frac{\omega}{4})} = \left(\frac{\omega}{2} \right)^{p_1} \left(\frac{4}{\omega} \right)^{p_1} = 2^{p_1}$$

is independent of ω.   □

Our next aim is to show that the conclusion of Lemma 5.3 holds in a full cylinder of the type $Q(\tau, \rho)$. The idea is to use the fact that at the time level

$$-\widehat{t} := t^* - \frac{\left(\frac{R}{2} \right)^2}{\psi \left(\frac{\omega}{4} \right)} \tag{5.13}$$

the function $v(x)$ is strictly below the level $1 - \frac{\omega}{4}$ in the cube $K_{\frac{R}{2}}$ and look at this time level as an initial condition to make the conclusion hold up to $t = 0$, eventually shrinking the cube. Again this is a sophisticated way of showing that somehow the equation behaves like the heat equation. As an intermediate step we need the following lemma.

Lemma 5.4. *Given $\nu_1 \in (0,1)$, there exists $s_1 \in \mathbb{N}$, depending only on the data, such that*

$$\left| \left\{ x \in K_{\frac{R}{4}} : v(x,t) > 1 - \frac{\omega}{2^{s_1}} \right\} \right| \le \nu_1 \left| K_{\frac{R}{4}} \right|, \quad \forall t \in (-\widehat{t}, 0).$$

Proof. We use a logarithmic estimate for the function $(v - k)_+$ in the cylinder $Q(\widehat{t}, \frac{R}{2})$, with the choices

$$k = 1 - \frac{\omega}{4} \quad \text{and} \quad c = \frac{\omega}{2^{n+2}},$$

where $n \in \mathbb{N}$ will be chosen later, as parameters in the standard logarithmic function (2.7). We have

$$v - k \le H_{v,k}^+ = \operatorname*{ess\,sup}_{Q(\widehat{t}, \frac{R}{2})} \left| \left(v - 1 + \frac{\omega}{4} \right)_+ \right| \le \frac{\omega}{4}. \tag{5.14}$$

If $H_{v,k}^+ \le \frac{\omega}{8}$, the result is trivial for the choice $s_1 = 3$. Assuming $H_{v,k}^+ > \frac{\omega}{8}$, recall from section 2.3 that the logarithmic function $\psi^+(v)$ is defined in the whole domain of v, $Q(\widehat{t}, \frac{R}{2})$ (since it is obvious that $H_{v,k}^+ - v + k + c > 0$), and given by

$$\psi^+_{\{H^+_{v,k},k,\frac{\omega}{2^{n+2}}\}}(v) = \begin{cases} \ln\left\{\dfrac{H^+_{v,k}}{H^+_{v,k} - v + k + \frac{\omega}{2^{n+2}}}\right\} & \text{if } v > k + \frac{\omega}{2^{n+2}} \\ \\ 0 & \text{if } v \le k + \frac{\omega}{2^{n+2}}. \end{cases}$$

From (5.14), we can easily estimate

$$\psi^+(v) \le n \ln 2 \qquad \text{since} \qquad \frac{H^+_{v,k}}{H^+_{v,k} - v + k + \frac{\omega}{2^{n+2}}} \le \frac{\frac{\omega}{4}}{\frac{\omega}{2^{n+2}}} = 2^n$$

and the derivative (here in the nonvanishing case $v > k + c$)

$$\left|(\psi^+)'(v)\right| = \left|\frac{-1}{H^+_{v,k} - v + k + c}\right| \le \left|\frac{1}{c}\right| = \frac{2^{n+2}}{\omega}.$$

To obtain the estimate, choose a cutoff function $0 < \zeta(x) \le 1$, defined on $K_{\frac{R}{2}}$ and such that

$$\zeta = 1 \text{ in } K_{\frac{R}{4}} \qquad \text{and} \qquad |\nabla\zeta| \le \frac{C}{R},$$

and multiply (5.2) by $2\psi^+(v_h)(\psi^+)'(v_h)\zeta^2$. Integrating in time in $(-\hat{t}, t)$, with $t \in (-\hat{t}, 0)$, and performing the usual integrations by parts and passages to the limit in h for the Steklov averages, we obtain, concerning the time part,

$$\int_{K_{\frac{R}{2}}\times\{t\}} \left[\psi^+(v)\right]^2 \zeta^2 - \int_{K_{\frac{R}{2}}\times\{-\hat{t}\}} \left[\psi^+(v)\right]^2 \zeta^2.$$

Now observe that as a consequence of Lemma 5.3, we have $v(x, -\hat{t}) < k$ in the cube $K_{\frac{R}{2}}$, which implies that

$$\left[\psi^+(v)\right](x, -\hat{t}) = 0, \quad x \in K_{\frac{R}{2}}.$$

As for the space part, we start by passing to the limit in h, thus getting

$$\int_{-\hat{t}}^{t}\int_{K_{\frac{R}{2}}} \alpha(v)\nabla v \cdot \nabla\left\{2\psi^+(v)(\psi^+)'(v)\zeta^2\right\}$$

$$= \int_{-\hat{t}}^{t}\int_{K_{\frac{R}{2}}} \alpha(v)|\nabla v|^2\left\{2\left(1 + \psi^+(v)\right)\left[(\psi^+)'(v)\right]^2 \zeta^2\right\}$$

$$+ 2\int_{-\hat{t}}^{t}\int_{K_{\frac{R}{2}}} \alpha(v)\nabla v \cdot \nabla\zeta\left\{2\psi^+(v)(\psi^+)'(v)\zeta\right\} = (*).$$

Using Young's inequality, we estimate the second term as

$$\left| 2 \int_{-\hat{t}}^{t} \int_{K_{\frac{R}{2}}} \alpha(v) \nabla v \cdot \nabla \zeta \left\{ 2 \psi^+(v) \left(\psi^+ \right)'(v) \, \zeta \right\} \right|$$

$$\leq 2 \int_{-\hat{t}}^{t} \int_{K_{\frac{R}{2}}} \alpha(v) |\nabla v|^2 \zeta^2 \psi^+(v) \left[(\psi^+)'(v) \right]^2 + 2 \int_{-\hat{t}}^{t} \int_{K_{\frac{R}{2}}} \alpha(v) |\nabla \zeta|^2 \psi^+(v),$$

and as a consequence we get

$$(*) \geq 2 \int_{-\hat{t}}^{t} \int_{K_{\frac{R}{2}}} \alpha(v) |\nabla v|^2 \zeta^2 \left[(\psi^+)'(v) \right]^2 - 2 \int_{-\hat{t}}^{t} \int_{K_{\frac{R}{2}}} \alpha(v) |\nabla \zeta|^2 \psi^+(v)$$

$$\geq -2n \ln 2 \, \frac{C}{R^2} \int_{-\hat{t}}^{t} \int_{K_{\frac{R}{2}}} \alpha(v) \chi_{\{v > 1 - \frac{\omega}{4}\}} \geq -2n \ln 2 \, \frac{C}{R^2} \, \hat{t} \left| K_{\frac{R}{2}} \right| C_1 \, \psi \left(\frac{\omega}{4} \right),$$

because, in the relevant set,

$$\alpha(v) \leq C_1 \, \psi(1 - v) \leq C_1 \, \psi \left(1 - \left(1 - \frac{\omega}{4} \right) \right) = C_1 \, \psi \left(\frac{\omega}{4} \right).$$

Now we observe that, due to our choice of t^*,

$$\hat{t} = -t^* + \frac{\left(\frac{R}{2} \right)^2}{\psi \left(\frac{\omega}{4} \right)} \leq \theta R^2,$$

and so we can condensate the two estimates in the logarithmic estimate

$$\sup_{-\hat{t} \leq t \leq 0} \int_{K_{\frac{R}{2}} \times \{t\}} \left[\psi^+(v) \right]^2 \zeta^2 \leq C \, n \, \theta \left| K_{\frac{R}{2}} \right| \psi \left(\frac{\omega}{4} \right). \qquad (5.15)$$

Now, since the integrand is nonnegative, we estimate below the left hand side of (5.15) integrating over the smaller set

$$S = \left\{ x \in K_{\frac{R}{4}} : v(x,t) > 1 - \frac{\omega}{2^{n+2}} \right\} \subset K_{\frac{R}{2}}.$$

Observing that in S, $\zeta = 1$ and

$$\left[\psi^+(v) \right]^2 \geq \left[\ln \left(2^{n-1} \right) \right]^2 = (n-1)^2 (\ln 2)^2,$$

we get

$$(n-1)^2 (\ln 2)^2 |S| \leq C \, n \, \theta \left| K_{\frac{R}{2}} \right| \psi \left(\frac{\omega}{4} \right)$$

and consequently,

$$\left| \left\{ x \in K_{\frac{R}{4}} : v(x,t) > 1 - \frac{\omega}{2^{n+2}} \right\} \right| \leq C \, \frac{n}{(n-1)^2} \, \frac{\psi \left(\frac{\omega}{4} \right)}{\phi \left(\frac{\omega}{2^m} \right)} \left| K_{\frac{R}{4}} \right|.$$

To prove the lemma we just need to choose

$$s_1 = n + 2 \qquad \text{with} \qquad n \geq 1 + \frac{2C}{\nu_1} \frac{\psi\left(\frac{\omega}{4}\right)}{\phi\left(\frac{\omega}{2^m}\right)},$$

since if $n \geq 1 + \frac{2}{\alpha}$ then

$$\frac{n}{(n-1)^2} \leq \alpha, \quad \alpha > 0.$$

Concerning the independence of ω, observe that

$$\frac{\psi\left(\frac{\omega}{4}\right)}{\phi\left(\frac{\omega}{2^m}\right)} = \frac{\left(\frac{\omega}{4}\right)^{p_1}}{\left(\frac{\omega}{2^m}\right)^{p_2}} = 2^{mp_2 - p_1} \omega^{p_1 - p_2} \leq 2^{mp_2 - p_1},$$

because $p_1 > p_2$ and $\omega < 1$, and this expression does not depended on ω. We strongly emphasize that it is crucial at this point that both degeneracies are powers and that $p_1 > p_2$. $\qquad \square$

We now arrive at the main result of this section that establishes the first alternative.

Proposition 5.5. *There exists a constant $s_1 \in \mathbb{N}$, depending only on the data, such that if (5.7) holds then*

$$v(x,t) < 1 - \frac{\omega}{2^{s_1+1}}, \quad \text{a.e. in } Q\left(\widehat{t}, \frac{R}{8}\right).$$

Proof. Consider the cylinder for which (5.7) holds, let

$$R_n = \frac{R}{8} + \frac{R}{2^{n+3}}, \quad n = 0, 1, \ldots$$

and construct the family of nested and shrinking cylinders $Q(\widehat{t}, R_n)$, where $\widehat{t}$ is given by (5.13). Take piecewise smooth cutoff functions $0 < \zeta_n(x) \leq 1$, not depending on t, defined in K_{R_n} and satisfying the assumptions

$$\zeta_n = 1 \text{ in } K_{R_{n+1}}, \quad |\nabla \zeta_n| \leq \frac{2^{n+4}}{R} \quad \text{and} \quad |\Delta \zeta_n| \leq \frac{2^{2(n+4)}}{R^2}.$$

Take also

$$k_n = 1 - \frac{\omega}{2^{s_1+1}} - \frac{\omega}{2^{s_1+1+n}}, \quad n = 0, 1, \ldots,$$

(where $s_1 > 1$ is to be chosen) and derive local energy inequalities similar to those obtained in Lemma 5.3, now for the functions

$$(v_\omega - k_n)_+^2 \zeta_n^2, \quad \text{with} \quad v_\omega := \min\left(v, 1 - \frac{\omega}{2^{s_1}}\right)$$

and in the cylinders $Q\left(\widehat{t}, R_n\right)$. Observing that, due to Lemma 5.3, we have

$$v(x, -\widehat{t}) < 1 - \frac{\omega}{4} \le k_n \quad \text{in the cube } K_{\frac{R}{2}} \supset K_{R_n},$$

which implies that

$$(v - k_n)_+(x, -\widehat{t}) = 0, \quad x \in K_{R_n}, \quad n = 0, 1, \ldots.$$

Performing the same type of reasoning used in Lemma 5.3 (which we shall not repeat here), we get

$$\sup_{-\widehat{t}<t<0} \int_{K_{R_n} \times \{t\}} (v_\omega - k_n)_+^2 \zeta_n^2 + \psi\left(\frac{\omega}{2^{s_1}}\right) \int_{-\widehat{t}}^0 \int_{K_{R_n}} |\zeta_n \nabla (v_\omega - k_n)_+|^2$$

$$\le \left\{\frac{4C_1{}^2 \, \psi\left(\frac{\omega}{2^{s_1-1}}\right)^2}{(2C_0 - 1)\, \psi\left(\frac{\omega}{2^{s_1}}\right)} + 8\,\psi\left(\frac{\omega}{2^{s_1}}\right)\right\} \frac{2^{2(n+4)}}{R^2} \left(\frac{\omega}{2^{s_1}}\right)^2 \int_{-\widehat{t}}^0 \int_{K_{R_n}} \chi_{\{v_\omega \ge k_n\}}.$$

The change in the time variable $\overline{t} = \left(\frac{R}{2}\right)^2 \frac{t}{\widehat{t}}$, with the new function

$$\overline{v_\omega}(\cdot, \overline{t}) = v_\omega(\cdot, t),$$

leads to

$$\frac{\left(\frac{R}{2}\right)^2}{\psi\left(\frac{\omega}{2^{s_1}}\right) \widehat{t}} \sup_{-\left(\frac{R}{2}\right)^2<t<0} \int_{K_{R_n} \times \{t\}} (\overline{v_\omega} - k_n)_+^2 \zeta_n^2$$

$$+ \int_{-\left(\frac{R}{2}\right)^2}^0 \int_{K_{R_n}} |\zeta_n \nabla (\overline{v_\omega} - k_n)_+|^2$$

$$\le \left\{\frac{4C_1{}^2 \, \psi\left(\frac{\omega}{2^{s_1-1}}\right)^2}{(2C_0 - 1)\, \psi\left(\frac{\omega}{2^{s_1}}\right)^2} + 8\right\} \frac{2^{2(n+4)}}{R^2} \left(\frac{\omega}{2^{s_1}}\right)^2 \int_{-\left(\frac{R}{2}\right)^2}^0 \int_{K_{R_n}} \chi_{\{\overline{v_\omega} \ge k_n\}},$$

after multiplication of the whole expression by $\dfrac{\left(\frac{R}{2}\right)^2}{\psi\left(\frac{\omega}{2^{s_1}}\right) \widehat{t}}$. Now, if s_1 is chosen

sufficiently large, we have

$$\frac{\left(\frac{R}{2}\right)^2}{\psi\left(\frac{\omega}{2^{s_1}}\right) \widehat{t}} \ge 1 \tag{5.16}$$

and obtain, with

$$\gamma = \left\{\frac{4C_1{}^2 \, \psi(\frac{\omega}{2^{s_1-1}})^2}{(2C_0 - 1)\, \psi(\frac{\omega}{2^{s_1}})^2} + 8\right\},$$

the simplified inequality

$$\|(\overline{v_\omega} - k_n)_+ \; \zeta_n\|^2_{V^2(Q((\frac{R}{2})^2, R_n))} \le \gamma \; \frac{2^{2(n+4)}}{R^2} \left(\frac{\omega}{2^{s_1}}\right)^2 \int_{-(\frac{R}{2})^2}^0 \int_{K_{R_n}} \chi_{\{\overline{v_\omega} \ge k_n\}}.$$

Define, for each n,

$$A_n = \int_{-(\frac{R}{2})^2}^0 \int_{K_{R_n}} \chi_{\{\overline{v_\omega} \ge k_n\}} \; dx \; d\bar{t}$$

and observe that the following estimates hold, by a reasoning similar to the one that led to (5.10):

$$\frac{1}{2^{2(n+2)}} \left(\frac{\omega}{2^{s_1}}\right)^2 A_{n+1} = |k_{n+1} - k_n|^2 \; A_{n+1}$$

$$\le \|(\overline{v_\omega} - k_n)_+\|^2_{2,Q((\frac{R}{2})^2, R_{n+1})}$$

$$\le \|(\overline{v_\omega} - k_n)_+ \zeta_n\|^2_{2,Q((\frac{R}{2})^2, R_n)}$$

$$\le C \; \|(\overline{v_\omega} - k_n)_+ \; \zeta_n\|^2_{V^2(Q((\frac{R}{2})^2, R_n))} \; A_n^{\frac{2}{d+2}}$$

$$\le \gamma \; \frac{2^{2(n+4)}}{R^2} \left(\frac{\omega}{2^{s_1}}\right)^2 A_n^{1+\frac{2}{d+2}}.$$

Next, define the numbers

$$X_n = \frac{A_n}{|Q\left((\frac{R}{2})^2, R_n\right)|},$$

divide the inequality by $\left|Q\left((\frac{R}{2})^2, R_{n+1}\right)\right|$ and obtain the recursive relation

$$X_{n+1} \le \gamma \, 4^{2n} \, X_n^{1+\frac{2}{d+2}}.$$

Using again Lemma 2.9 we conclude that if

$$X_0 \le \gamma^{-\frac{d+2}{2}} 4^{-\frac{(d+2)^2}{2}} =: \nu_1 \in (0,1) \tag{5.17}$$

then

$$X_n \longrightarrow 0. \tag{5.18}$$

Apply Lemma 5.4 with this ν_1 and conclude that there exists s_1, depending only on the data, such that

$$\left|\left\{x \in K_{\frac{R}{4}} : v(x,t) \ge 1 - \frac{\omega}{2^{s_1}}\right\}\right| \le \nu_1 \left|K_{\frac{R}{4}}\right|, \quad \forall t \in (-\hat{t}, 0),$$

and obtain (5.17) as follows:

$$
X_0 = \frac{\int_{-(\frac{R}{2})^2}^0 \int_{K_{\frac{R}{4}}} \chi_{\{\overline{v_\omega} \geq 1 - \frac{\omega}{2^{s_1}}\}}}{\left| Q\left((\frac{R}{2})^2, \frac{R}{4}\right)\right|}
$$

$$
= \frac{(\frac{R}{2})^2}{\widehat{t}} \frac{\int_{-\widehat{t}}^0 \int_{K_{\frac{R}{4}}} \chi_{\{v \geq 1 - \frac{\omega}{2^{s_1}}\}}}{\left| Q\left((\frac{R}{2})^2, \frac{R}{4}\right)\right|}
$$

$$
\leq \frac{(\frac{R}{2})^2}{\widehat{t}} \frac{\widehat{t}\left|\left\{x \in K_{\frac{R}{4}} : v(x,t) \geq 1 - \frac{\omega}{2^{s_1}}\right\}\right|}{(\frac{R}{2})^2 \left|K_{\frac{R}{4}}\right|} \leq \nu_1.
$$

Now the conclusion easily follows from (5.18). In fact, observe that

$$
R_n \searrow \frac{R}{8} \qquad \text{and} \qquad k_n \nearrow 1 - \frac{\omega}{2^{s_1+1}},
$$

and since (5.18) implies that $A_n \to 0$, we conclude that

$$
\left|\left\{(x,\overline{t}) \in Q\left(\left(\frac{R}{2}\right)^2, \frac{R}{8}\right) : \overline{v_\omega}(x,\overline{t}) \geq 1 - \frac{\omega}{2^{s_1+1}}\right\}\right|
$$

$$
= \left|\left\{(x,t) \in Q\left(\widehat{t}, \frac{R}{8}\right) : v(x,t) \geq 1 - \frac{\omega}{2^{s_1+1}}\right\}\right| = 0
$$

and the proposition is proved. □

Remark 5.6. It is very important to remark that ν_1 in the proof is independent of s_1, since otherwise the reasoning would be fallacious. In fact, we have

$$
\nu_1 = \gamma^{-\frac{d+2}{2}} 4^{-\frac{(d+2)^2}{2}}
$$

and with our assumption that ψ is a power, we have

$$
\gamma = \left\{\frac{4C_1^2}{2C_0 - 1}\left[\frac{\psi(\frac{\omega}{2^{s_1-1}})}{\psi(\frac{\omega}{2^{s_1}})}\right]^2 + 8\right\} = \left\{\frac{4C_1^2}{2C_0 - 1} 2^{2p_1} + 8\right\},
$$

which clearly shows what was claimed. This also proves the independence of s_1 with respect to ω, through Lemma 5.4, together with the remark that concerning the choice (5.16), we have

$$
\frac{(\frac{R}{2})^2}{\psi\left(\frac{\omega}{2^{s_1}}\right)\widehat{t}} \geq \frac{\phi\left(\frac{\omega}{2^m}\right)}{4\psi\left(\frac{\omega}{2^{s_1}}\right)} \qquad \text{because} \qquad \widehat{t} \leq \theta R^2
$$

and so it can be made larger than 1 if

$$\frac{\phi\left(\frac{\omega}{2^m}\right)}{4\psi\left(\frac{\omega}{2^{s_1}}\right)} = \frac{\left(\frac{\omega}{2^m}\right)^{p_2}}{4\left(\frac{\omega}{2^{s_1}}\right)^{p_1}} = \omega^{p_2-p_1}2^{p_1 s_1-mp_2-2} \geq 1.$$

Since $p_1 > p_2$ and $\omega < 1$, it is enough to choose

$$s_1 \geq \frac{2 + mp_2}{p_1},$$

that clearly does not depend on ω.

Corollary 5.7. *There exists a constant $\sigma_0 \in (0,1)$, depending only on the data, such that if (5.7) holds then*

$$\operatorname*{ess\,osc}_{Q(\hat{t},\frac{R}{8})} v \leq \sigma_0\,\omega. \tag{5.19}$$

Proof. We can use Proposition 5.5 to obtain $s_1 \in \mathbb{N}$ such that

$$\operatorname*{ess\,sup}_{Q(\hat{t},\frac{R}{8})} v \leq 1 - \frac{\omega}{2^{s_1+1}}$$

and from this we get

$$\operatorname*{ess\,osc}_{Q(\hat{t},\frac{R}{8})} v = \operatorname*{ess\,sup}_{Q(\hat{t},\frac{R}{8})} v - \operatorname*{ess\,inf}_{Q(\hat{t},\frac{R}{8})} v \leq 1 - \frac{\omega}{2^{s_1+1}} - 0 = \left(1 - \frac{1}{2^{s_1+1}}\right)\omega$$

and the corollary follows with $\sigma_0 = \left(1 - \frac{1}{2^{s_1+1}}\right)$.   □

5.4 Behaviour Near the other Degeneracy

Let us now suppose that (5.7) does not hold. In that case we have the complementary condition and, since $1 - \frac{\omega}{2} \geq \frac{\omega}{2}$, this means that for every cylinder of the type $Q_R^{t^*}$, we have

$$\left|\left\{(x,t) \in Q_R^{t^*} : v(x,t) < \frac{\omega}{2}\right\}\right| \leq (1 - \nu_0)\left|Q_R^{t^*}\right|. \tag{5.20}$$

We are in the case for which we have to analyze the behaviour of v near 0, the other point where $\alpha(v)$ degenerates. We will show that also in this case a conclusion similar to (5.19) can be reached. Recall that the constant ν_0 has already been determined in the previous section and is given by (5.11).

Fix a cylinder $Q_R^{t^*} \subset Q(\theta R^2, R)$ for which (5.20) holds. There exists a time level

$$t^\circ \in \left[t^* - \frac{R^2}{\psi\left(\frac{\omega}{4}\right)}, t^* - \frac{\nu_0}{2}\frac{R^2}{\psi\left(\frac{\omega}{4}\right)}\right]$$

such that

$$\left|\left\{x \in K_R : v(x,t^\circ) < \frac{\omega}{2}\right\}\right| < \left(\frac{1-\nu_0}{1-\frac{\nu_0}{2}}\right) |K_R|. \tag{5.21}$$

In fact, if this is not true, then

$$\left|\left\{x \in K_R : v(x,t^\circ) < \frac{\omega}{2}\right\}\right| \geq \int_{t^* - \frac{R^2}{\psi(\frac{\omega}{4})}}^{t^* - \frac{\nu_0}{2}\frac{R^2}{\psi(\frac{\omega}{4})}} \left|\left\{x \in K_R : v(x,t) < \frac{\omega}{2}\right\}\right| dt$$

$$\geq \left\{t^* - \frac{\nu_0}{2}\frac{R^2}{\psi\left(\frac{\omega}{4}\right)} - t^* + \frac{R^2}{\psi\left(\frac{\omega}{4}\right)}\right\}\left(\frac{1-\nu_0}{1-\frac{\nu_0}{2}}\right)|K_R|$$

$$= (1-\nu_0)\frac{R^2}{\psi\left(\frac{\omega}{4}\right)}|K_R|$$

$$= (1-\nu_0)|Q_R^{t^*}|,$$

which contradicts our assumption. The next lemma asserts that the set where $v(x)$ is close to its infimum is small, not only at a specific time level, but for all time levels near the top of the cylinder $Q_R^{t^*}$.

Lemma 5.8. *There exists $s_2 \in \mathbb{N}$, depending only on the data, such that*

$$\left|\left\{x \in K_R : v(x,t) < \frac{\omega}{2^{s_2}}\right\}\right| \leq \left(1-\left(\frac{\nu_0}{2}\right)^2\right)|K_R|,$$

for all $t \in \left[t^* - \frac{\nu_0}{2}\frac{R^2}{\psi(\frac{\omega}{4})}, t^*\right]$.

Proof. We use a logarithmic estimate for the function $(v-k)_-$ in the cylinder $K_R \times (t^\circ, t^*)$, with the choices

$$k = \frac{\omega}{2} \quad \text{and} \quad c = \frac{\omega}{2^{n+1}},$$

where $n \in \mathbb{N}$ will be chosen later, as parameters in the standard logarithmic function (2.7). We have

$$k - v \leq H_{v,k}^- = \operatorname*{ess\,sup}_{K_R \times (t^\circ,t^*)} \left|\left(v-\frac{\omega}{2}\right)_-\right| \leq \frac{\omega}{2}. \tag{5.22}$$

If $H_{v,k}^+ \leq \frac{\omega}{4}$, the result is trivial for the choice $s_2 = 2$. Assuming $H_{v,k}^+ > \frac{\omega}{4}$, recall from section 2.3 that the logarithmic function $\psi^-(v)$ is defined in the whole domain of v, $K_R \times (t^\circ, t^*)$ (since it is obvious that $H_{v,k}^- + v - k + c > 0$), and given by

$$\psi^-_{\{H_{v,k}^-,k,\frac{\omega}{2^{n+1}}\}}(v) = \begin{cases} \ln\left\{\dfrac{H_{v,k}^-}{H_{v,k}^- + v - k + \frac{\omega}{2^{n+1}}}\right\} & \text{if } v < k - \frac{\omega}{2^{n+1}} \\[2ex] 0 & \text{if } v \geq k - \frac{\omega}{2^{n+1}}. \end{cases}$$

From (5.22), we can easily estimate

$$\psi^-(v) \le n \ln 2 \qquad \text{since} \qquad \frac{H^-_{v,k}}{H^-_{v,k} + v - k + \frac{\omega}{2^{n+1}}} \le \frac{\frac{\omega}{2}}{\frac{\omega}{2^{n+1}}} = 2^n$$

and the derivative (here in the nonvanishing case $v < k - c$)

$$\left|\left(\psi^-\right)'(v)\right| = \left|\frac{-1}{H^-_{v,k} + v - k + c}\right| \le \left|\frac{1}{c}\right| = \frac{2^{n+1}}{\omega}.$$

Choose a piecewise smooth cutoff function $0 < \zeta(x) \le 1$, defined on K_R and such that, for a certain $\sigma \in (0,1)$,

$$\zeta = 1 \ \text{ in } \ K_{(1-\sigma)R} \qquad \text{and} \qquad |\nabla \zeta| \le \frac{C}{\sigma R},$$

and multiply (5.2) by $2\psi^-(v_h)\,(\psi^-)'(v_h)\,\zeta^2$. Integrating in time in (t°, t), with $t \in (t^\circ, t^*)$, performing the usual integrations by parts, and passing to the limit in h for the Steklov averages, we obtain, concerning the time part,

$$\int_{K_R \times \{t\}} \left[\psi^-(v)\right]^2 \zeta^2 - \int_{K_R \times \{t^\circ\}} \left[\psi^-(v)\right]^2 \zeta^2.$$

As for the space part, we start by passing to the limit in h, thus getting

$$\int_{t^\circ}^t \int_{K_R} \alpha(v) \nabla v \cdot \nabla \left\{2\psi^-(v)\,(\psi^-)'(v)\,\zeta^2\right\}$$

$$= \int_{t^\circ}^t \int_{K_R} \alpha(v) |\nabla v|^2 \left\{2\left(1 + \psi^-(v)\right)\left[(\psi^-)'(v)\right]^2 \zeta^2\right\}$$

$$+ 2 \int_{t^\circ}^t \int_{K_R} \alpha(v) \nabla v \cdot \nabla \zeta \left\{2\psi^-(v)\,(\psi^-)'(v)\,\zeta\right\} = (*).$$

Using Young's inequality to estimate the second term as in the proof of Lemma 5.4, we get

$$(*) \ge 2 \int_{t^\circ}^t \int_{K_R} \alpha(v) |\nabla v|^2 \zeta^2 \left[(\psi^-)'(v)\right]^2 - 2 \int_{t^\circ}^t \int_{K_R} \alpha(v) |\nabla \zeta|^2 \psi^-(v)$$

$$\ge -2n \ln 2 \frac{C}{\sigma^2 R^2} \int_{t^\circ}^t \int_{K_R} \alpha(v) \chi_{\{v < \frac{\omega}{2}\}} \ge -2n \ln 2 \frac{C}{\sigma^2 R^2} (t^* - t^\circ) |K_R| \, C_1 \phi\left(\frac{\omega}{2}\right),$$

because in the relevant set

$$\alpha(v) \le C_1 \phi(v) \le C_1 \phi\left(\frac{\omega}{2}\right).$$

Now we observe that

$$t^* - t^\circ \le t^* - t^* + \frac{R^2}{\psi\left(\frac{\omega}{4}\right)} = \frac{R^2}{\psi\left(\frac{\omega}{4}\right)},$$

and so we can condensate the two estimates in the logarithmic estimate

$$\sup_{t^\circ \le t \le t^*} \int_{K_R \times \{t\}} \left[\psi^-(v)\right]^2 \zeta^2 \le \int_{K_R \times \{t^\circ\}} \left[\psi^-(v)\right]^2 \zeta^2 + C\, n\, \frac{1}{\sigma^2} \frac{\phi\left(\frac{\omega}{2}\right)}{\psi\left(\frac{\omega}{4}\right)} |K_R|$$

$$\le C\, n^2 \left(\frac{1 - \nu_0}{1 - \frac{\nu_0}{2}}\right) |K_R| + C\, n\, \frac{1}{\sigma^2} \frac{\phi\left(\frac{\omega}{2}\right)}{\psi\left(\frac{\omega}{4}\right)} |K_R|,$$

because $\psi^-(v) = 0$ in the set $\{v > k = \frac{\omega}{2}\}$ and using (5.21). Now, since the integrand is nonnegative, we estimate below the left hand side in the expression above integrating over the smaller set

$$S = \left\{x \in K_{(1-\sigma)R} : v(x,t) < \frac{\omega}{2^{n+1}}\right\} \subset K_R, \quad t \in (t^\circ, t^*).$$

Observing that in S, $\zeta = 1$ and

$$\left[\psi^-(v)\right]^2 \ge \left[\ln\left(2^{n-1}\right)\right]^2 = (n-1)^2 (\ln 2)^2,$$

we get

$$|S| \le C \left(\frac{n}{n-1}\right)^2 \left(\frac{1-\nu_0}{1-\frac{\nu_0}{2}}\right) |K_R| + C\, \frac{n}{(n-1)^2}\, \frac{1}{\sigma^2} \frac{\phi\left(\frac{\omega}{2}\right)}{\psi\left(\frac{\omega}{4}\right)} |K_R|$$

$$\le C \left(\frac{n}{n-1}\right)^2 \left(\frac{1-\nu_0}{1-\frac{\nu_0}{2}}\right) |K_R| + C\, \frac{1}{n}\, \frac{1}{\sigma^2} \frac{\phi\left(\frac{\omega}{2}\right)}{\psi\left(\frac{\omega}{4}\right)} |K_R|.$$

On the other hand, we have

$$\left|\left\{x \in K_R : v(x,t) < \frac{\omega}{2^{n+1}}\right\}\right|$$

$$\le \left|\left\{x \in K_{(1-\sigma)R} : v(x,t) < \frac{\omega}{2^{n+1}}\right\}\right| + \left|K_R \setminus K_{(1-\sigma)R}\right|$$

$$\le |S| + d\sigma |K_R|,$$

so the conclusion is that for all $t \in (t^\circ, t^*)$,

$$\left|\left\{x \in K_R : v(x,t) < \frac{\omega}{2^{n+1}}\right\}\right|$$

$$\le C \left\{\left(\frac{n}{n-1}\right)^2 \left(\frac{1-\nu_0}{1-\frac{\nu_0}{2}}\right) + \frac{1}{n}\, \frac{1}{\sigma^2} \frac{\phi\left(\frac{\omega}{2}\right)}{\psi\left(\frac{\omega}{4}\right)} + d\sigma\right\} |K_R|.$$

Now, choose σ so small that $Cd\sigma \leq \frac{3}{8}\nu_0^2$ and n so large that

$$C\left(\frac{n}{n-1}\right)^2 \leq \left(1 - \frac{\nu_0}{2}\right)(1+\nu_0) =: \beta > 1 \quad \text{and} \quad \frac{C}{n\sigma^2}\frac{\phi\left(\frac{\omega}{2}\right)}{\psi\left(\frac{\omega}{4}\right)} \leq \frac{3}{8}\nu_0^2$$

and the lemma follows with $s_2 = n+1$.

Concerning the independence of s_2 with respect to ω, and since ν_0 has already been chosen respecting that independence, it is enough to note that, due to (5.6),

$$\frac{\phi\left(\frac{\omega}{2}\right)}{\psi\left(\frac{\omega}{4}\right)} \leq \frac{\phi\left(\frac{\omega}{2}\right)}{\phi\left(\frac{\omega}{2^m}\right)} = 2^{(m-1)p_2},$$

expression does not depended on ω.   □

Now we want to show that the same type of conclusion holds in an upper portion of the full cylinder $Q(\theta R^2, R)$, say for all $t \in \left(-\frac{\theta}{2}R^2, 0\right)$. We just have to use the fact that (5.20) holds for *all* cylinders of the type $Q_R^{t^*}$ and so, recalling (5.5), the conclusion of the previous lemma holds true for all time levels

$$t \geq \frac{R^2}{\psi\left(\frac{\omega}{4}\right)} - \frac{R^2}{\phi\left(\frac{\omega}{2^m}\right)} - \frac{\nu_0}{2}\frac{R^2}{\psi\left(\frac{\omega}{4}\right)}.$$

Using assumption (5.6), we have

$$\frac{R^2}{\psi\left(\frac{\omega}{4}\right)} - \frac{R^2}{\phi\left(\frac{\omega}{2^m}\right)} - \frac{\nu_0}{2}\frac{R^2}{\psi\left(\frac{\omega}{4}\right)} \leq \left\{\frac{1}{2}\left(1 - \frac{\nu_0}{2}\right) - 1\right\}\frac{R^2}{\phi\left(\frac{\omega}{2^m}\right)}$$

$$= -\left\{1 + \frac{\nu_0}{2}\right\}\frac{R^2}{2\phi\left(\frac{\omega}{2^m}\right)}$$

$$\leq -\frac{\theta}{2}R^2.$$

Corollary 5.9. *For all* $t \in \left(-\frac{\theta}{2}R^2, 0\right)$,

$$\left|\left\{x \in K_R : v(x,t) < \frac{\omega}{2^{s_2}}\right\}\right| \leq \left(1 - \left(\frac{\nu_0}{2}\right)^2\right)|K_R|.$$

The previous result means that the set of degeneracy at 0 does not fill the entire cube K_R, for all $t \in \left(-\frac{\theta}{2}R^2, 0\right)$. The next result uses this stability information.

Proposition 5.10. *For every* $\lambda_0 \in (0,1)$, *there exist constants* $s_2 < s_3 \in \mathbb{N}$ *and* $m_0 \in \mathbb{N}$, *depending only on the data, such that, if* $m > m_0$ *then*

$$\left|\left\{(x,t) \in Q\left(\frac{\theta}{2}R^2, R\right) : v(x,t) < \frac{\omega}{2^{s_3}}\right\}\right| \leq \lambda_0 \left|Q\left(\frac{\theta}{2}R^2, R\right)\right|.$$

Instead of proving this proposition directly for v we will formulate an equivalent result for a new unknown function defined by

$$w = A(v) := \int_0^v \alpha(s)\, ds, \quad v \in [0, \delta_0].$$

We find that w satisfies the equation

$$[B(w)]_t - \Delta w = 0, \quad \text{in } \mathcal{D}'(\Omega_T), \tag{5.23}$$

where $B = A^{-1}$ is the inverse of A; both functions are monotone increasing. Define also

$$\bar{\omega} := \int_0^\omega \alpha(s)\, ds,$$

and rephrase Corollary 5.9 in terms of w as follows.

Corollary 5.11. *There exists a number $s_4 \in \mathbb{N}$, depending only on the data, such that, for all $t \in \left(-\frac{\theta}{2}R^2, 0\right)$,*

$$\left| \left\{ x \in K_R : w(x,t) < \frac{\bar{\omega}}{2^{s_4}} \right\} \right| \le \left(1 - \left(\frac{\nu_0}{2} \right)^2 \right) |K_R|.$$

This fact is a consequence of the following simple reasoning:

$$v < \frac{\omega}{2^{s_2}} \iff w < \int_0^{\frac{\bar{\omega}}{2^{s_2}}} \alpha(s)\, ds = \int_0^{\frac{\bar{\omega}}{2^{s_2}}} s^{p_2}\, ds$$

$$= \frac{\left(\frac{\omega}{2^{s_2}} \right)^{p_2+1}}{p_2 + 1} = \frac{\int_0^\omega \alpha(s)\, ds}{2^{s_2(p_2+1)}}$$

$$= \frac{\bar{\omega}}{2^{s_4}}, \quad \text{with} \quad s_4 = s_2(p_2 + 1).$$

Proposition 5.12. *For every $\lambda_0 \in (0,1)$, there exist constants $s_5 \in \mathbb{N}$ and $m_0 \in \mathbb{N}$, depending only on the data, such that, if $m > m_0$ then*

$$\left| \left\{ (x,t) \in Q\left(\frac{\theta}{2}R^2, R \right) : w(x,t) < \frac{\bar{\omega}}{2^{s_5}} \right\} \right| \le \lambda_0 \left| Q\left(\frac{\theta}{2}R^2, R \right) \right|.$$

Proof. We derive an energy inequality from equation (5.23) and use again the iteration technique. To be entirely precise, we should argue with the Steklov averages, as before, but to simplify we only do the formal computations this time. Consider a piecewise smooth cutoff function $0 \le \zeta \le 1$, defined in $Q(\theta R^2, 2R)$, and satisfying the following set of assumptions

$$\zeta = 1 \text{ in } Q(\tfrac{\theta}{2}R^2, R); \qquad \zeta = 0 \text{ on } \partial_p Q(\theta R^2, 2R);$$

$$|\nabla \zeta| \le \frac{1}{R}; \qquad 0 \le \zeta_t \le C\,\frac{\phi(\frac{\omega}{2^m})}{R^2}.$$

Let
$$k = \frac{\overline{\omega}}{2^l}, \quad (l > s_4 \text{ to be chosen}),$$

multiply (5.23) by $-(w-k)_- \zeta^2$ and integrate in time over $(-\theta R^2, 0)$. Concerning the time part, we get (formally)

$$\int_{-\theta R^2}^0 \int_{K_{2R}} [B(w)]_t \ \left[-(w-k)_- \zeta^2\right]$$

$$= \int_{K_{2R}} \int_{-\theta R^2}^0 \left(\int_0^{(w-k)_-} B'(k-s)s\, ds\right)_t \zeta^2$$

$$= \int_{K_{2R} \times \{0\}} \left(\int_0^{(w-k)_-} B'(k-s)s\, ds\right) \zeta^2$$

$$- \int_{K_{2R} \times \{-\theta R^2\}} \left(\int_0^{(w-k)_-} B'(k-s)s\, ds\right) \zeta^2$$

$$-2 \int_{-\theta R^2}^0 \int_{K_{2R}} \left(\int_0^{(w-k)_-} B'(k-s)s\, ds\right) \zeta \zeta_t$$

$$\geq -C \frac{\overline{\omega}}{2^l} B\left(\frac{\overline{\omega}}{2^l}\right) \frac{\phi(\frac{\overline{\omega}}{2^m})}{R^2} \left|Q\left(\frac{\theta}{2} R^2, R\right)\right|,$$

as a consequence of the following three facts: $\zeta(x, -\theta R^2) = 0$, due to our choice of ζ,

$$\int_0^{(w-k)_-} B'(k-s)s\, ds \geq 0,$$

since $B' \geq 0$, and

$$\int_0^{(w-k)_-} B'(k-s)s\, ds \leq (w-k)_- \int_0^{(w-k)_-} B'(k-s)\, ds$$

$$= (w-k)_- \left[-B\left(k-(w-k)_-\right) + B(k)\right]$$

$$= (w-k)_- \left[B(k) - B(w)\right]$$

$$\leq kB(k) = \frac{\overline{\omega}}{2^l} B\left(\frac{\overline{\omega}}{2^l}\right).$$

From the space part, comes

$$\int_{-\theta R^2}^0 \int_{K_{2R}} \nabla w \cdot \nabla \left[-(w-k)_- \zeta^2\right] = \int_{-\theta R^2}^0 \int_{K_{2R}} |\zeta \nabla(w-k)_-|^2$$

$$-2 \int_{-\theta R^2}^0 \int_{K_{2R}} (w-k)_- \zeta \nabla w \cdot \nabla \zeta$$

$$\geq \frac{1}{2} \int_{-\theta R^2}^0 \int_{K_{2R}} |\zeta \nabla(w-k)_-|^2 - C \left(\frac{\overline{\omega}}{2^l}\right)^2 \frac{1}{R^2} \left|Q\left(\frac{\theta}{2} R^2, R\right)\right|,$$

and we combine both estimates in

$$\int_{-\theta R^2}^{0} \int_{K_{2R}} |\zeta \nabla (w-k)_-|^2$$

$$\leq C \frac{\overline{\omega}}{2^l} B\left(\frac{\overline{\omega}}{2^l}\right) \frac{\phi\left(\frac{\omega}{2^m}\right)}{R^2} \left|Q\left(\frac{\theta}{2}R^2, R\right)\right|$$

$$+ C \left(\frac{\overline{\omega}}{2^l}\right)^2 \frac{1}{R^2} \left|Q\left(\frac{\theta}{2}R^2, R\right)\right|.$$

Integrating over the smaller set $Q\left(\frac{\theta}{2}R^2, R\right)$, where $\zeta = 1$, we arrive at

$$\|\nabla(w-k)_-\|^2_{2,Q\left(\frac{\theta}{2}R^2,R\right)}$$

$$\leq \left\{1 + \frac{2^l}{\overline{\omega}} B\left(\frac{\overline{\omega}}{2^l}\right) \phi\left(\frac{\omega}{2^m}\right)\right\} \frac{C}{R^2} \left(\frac{\overline{\omega}}{2^l}\right)^2 \left|Q\left(\frac{\theta}{2}R^2, R\right)\right|. \qquad (5.24)$$

Let us now use Lemma 2.7, with the function $w(x,t)$ defined for $t \in [-\frac{\theta}{2}R^2, 0]$, and the levels

$$k_1 = \frac{\overline{\omega}}{2^{l+1}} \quad \text{and} \quad k_2 = k = \frac{\overline{\omega}}{2^l} \; ; \; l = s_4, s_4 + 1, \ldots$$

We know from Corollary 5.11, and since $l \geq s_4$, that

$$\left|\left\{x \in K_R : w(x,t) > \frac{\overline{\omega}}{2^l}\right\}\right| \geq |K_R| - \left|\left\{x \in K_R : w(x,t) < \frac{\overline{\omega}}{2^{s_4}}\right\}\right|$$

$$\geq |K_R| - \left(1 - \left(\frac{\nu_0}{2}\right)^2\right) |K_R|$$

$$= \left(\frac{\nu_0}{2}\right)^2 |K_R|,$$

for all $t \in [-\frac{\theta}{2}R^2, 0]$. Next, define

$$A_l(t) = \left\{x \in K_R : w(x,t) < \frac{\overline{\omega}}{2^l}\right\} \quad \text{and} \quad A_l = \int_{-\frac{\theta}{2}R^2}^{0} |A_l(t)| \; dt,$$

and using Lemma 2.7, obtain

$$\left(\frac{\overline{\omega}}{2^l} - \frac{\overline{\omega}}{2^{l+1}}\right) |A_{l+1}(t)| \leq \frac{C \, R^{d+1}}{\left(\frac{\nu_0}{2}\right)^2 |K_R|} \int_{A_l(t) \setminus A_{l+1}(t)} |\nabla w|.$$

Integrate this inequality in time over $[-\frac{\theta}{2}R^2, 0]$, use Hölder's inequality and square both sides, to get

$$\left(\frac{\overline{\omega}}{2^l}\right)^2 A_{l+1}^2 \leq C \frac{R^2}{\nu_0^4} (A_l - A_{l+1}) \int_{-\frac{\theta}{2}R^2}^0 \int_{A_l(t)\setminus A_{l+1}(t)} |\nabla w|^2$$

$$\leq C \frac{R^2}{\nu_0^4} (A_l - A_{l+1}) \int_{-\frac{\theta}{2}R^2}^0 \int_{K_R} |\nabla(w-k)_-|^2$$

$$\leq \frac{C}{\nu_0^4} (A_l - A_{l+1}) \left\{ 1 + \frac{2^l}{\overline{\omega}} B\left(\frac{\overline{\omega}}{2^l}\right) \phi\left(\frac{\omega}{2^m}\right) \right\} \left(\frac{\overline{\omega}}{2^l}\right)^2 \left| Q\left(\frac{\theta}{2}R^2, R\right) \right|$$

using also inequality (5.24). Adding these inequalities for

$$l = s_4, s_4 + 1, \ldots, s_5 - 1,$$

where s_5 is to be chosen, we get

$$\sum_{l=s_4}^{s_5-1} A_{l+1}^2 \leq \frac{C}{\nu_0^4} (A_{s_4} - A_{s_5}) \left\{ 1 + \frac{2^{s_5}}{\overline{\omega}} B\left(\frac{\overline{\omega}}{2^{s_4}}\right) \phi\left(\frac{\omega}{2^m}\right) \right\} \left| Q\left(\frac{\theta}{2}R^2, R\right) \right|,$$

and since $A_{s_4} - A_{s_5} \leq \left| Q(\frac{\theta}{2}R^2, R) \right|$ and

$$\sum_{l=s_4}^{s_5-1} A_{l+1}^2 \geq (s_5 - s_4) A_{s_5}^2,$$

we finally conclude that

$$A_{s_5} \leq \frac{C}{\nu_0^2} (s_5 - s_4)^{-\frac{1}{2}} \left\{ 1 + \frac{2^{s_5}}{\overline{\omega}} B\left(\frac{\overline{\omega}}{2^{s_4}}\right) \phi\left(\frac{\omega}{2^m}\right) \right\}^{\frac{1}{2}} \left| Q\left(\frac{\theta}{2}R^2, R\right) \right|.$$

To prove the result, we choose s_5 so large that

$$\frac{C}{\nu_0^2} (s_5 - s_4)^{-\frac{1}{2}} \leq \frac{\lambda_0}{\sqrt{2}}$$

and m_0 so large that

$$\frac{2^{s_5}}{\overline{\omega}} B\left(\frac{\overline{\omega}}{2^{s_4}}\right) \phi\left(\frac{\omega}{2^{m_0}}\right) \leq 1. \tag{5.25}$$

Concerning the dependence on ω, we immediately conclude that, since ν_0 and s_4 are independent of ω (Lemma 5.3 and Corollary 5.11), so is s_5. The independence of m_0 on ω is more delicate and again crucially related to the fact that ϕ is a power. Observe that as a consequence of this fact

$$B(s) = \{(p_2+1)s\}^{\frac{1}{p_2+1}} \qquad \text{and} \qquad \overline{\omega} = \frac{\omega^{p_2+1}}{p_2+1}.$$

So, from (5.25), we must choose m_0 such that

$$\left(\frac{\omega}{2^{m_0}}\right)^{p_2} \leq \frac{\omega^{p_2+1}}{(p_2+1)2^{s_5}} \frac{2^{\frac{s_4}{p_2+1}}}{\omega}$$

that is

$$2^{m_0 p_2} \geq (p_2+1) \, 2^{s_5 - \frac{s_4}{p_2+1}},$$

and now it is clear that m_0 can be chosen independently of ω. □

We can now obtain Proposition 5.10 by rephrasing the contents of Proposition 5.12. In fact,

$$w(x,t) < \frac{\overline{\omega}}{2^{s_5}} \iff v(x,t) < \frac{\omega}{2^{\frac{s_5}{p_2+1}}}$$

and we conclude putting $s_3 = \frac{s_5}{p_2+1}$.

We next explain how to choose the constant m (independently of ω) and consequently fix the height of the cylinder $Q(\theta R^2, R)$. We follow closely the idea in [13]. Let

$$\phi_1 = \phi\left(\frac{\omega}{2^{s_3}}\right), \quad \phi_2 = \phi\left(\frac{\omega}{2^{s_3+2}}\right), \quad \mu = \phi_2\left(\frac{\phi_1}{\phi_2}\right)^{\frac{d+2}{2}},$$

and choose $m > m_0$ as the smallest real number such that

$$\frac{\mu}{\phi\left(\frac{\omega}{2^m}\right)} = n_0, \quad \text{for some integer } n_0 \in \mathbb{N}. \tag{5.26}$$

Since ϕ is a power, it is clear that m is independent of ω.

Then break $Q(\theta R^2, R)$ again, this time into n_0 subcylinders of the form

$$Q_R^j = K_R \times \left(-j\frac{R^2}{\mu}, -(j-1)\frac{R^2}{\mu}\right) \, ; \quad j = 1, 2, \ldots, n_0.$$

Since these cylinders are disjoint and they exhaust $Q(\theta R^2, R)$, from Proposition 5.10 it follows that

$$\exists j_0 \in \{1, \ldots, n_0\} \, : \, \left|\left\{(x,t) \in Q_R^{j_0} : v(x,t) < \frac{\omega}{2^{s_3}}\right\}\right| \leq \lambda_0 \left|Q_R^{j_0}\right|. \tag{5.27}$$

We now use this information to show that in a smaller cylinder the function v is strictly away from the degeneracy at 0.

Lemma 5.13. *The number s_3 can be chosen such that*

$$v(x,t) > \frac{\omega}{2^{s_3+1}}, \quad \text{a.e. in } Q_{\frac{R}{2}}^{j_0}.$$

Proof. Let $v_\omega := \max\{v, \frac{\omega}{2^{s_3+2}}\}$. Define

$$R_n = \frac{R}{2} + \frac{R}{2^{n+1}}, \quad n = 0, 1, \ldots,$$

and construct the family of nested and shrinking cylinders

$$Q_{R_n}^{j_0} = K_{R_n} \times \left(-j_0 \frac{R_n^2}{\mu}, -(j_0 - 1)\frac{R_n^2}{\mu}\right).$$

Consider piecewise smooth cutoff functions $0 \leq \zeta_n \leq 1$, defined in these cylinders, and satisfying the following set of assumptions

$$\zeta_n = 1 \text{ in } Q_{R_{n+1}}^{j_0}; \qquad\qquad \zeta_n = 0 \text{ on } \partial_p Q_{R_n}^{j_0};$$

$$|\nabla \zeta_n| \leq \frac{2^{n+1}}{R};$$

$$|\Delta \zeta_n| \leq \frac{2^{2(n+1)}}{R^2}; \qquad\qquad 0 \leq (\zeta_n)_t \leq 2^{2(n+1)}\frac{\mu}{R^2}.$$

Let

$$k_n = \frac{\omega}{2^{s_3+1}} + \frac{\omega}{2^{s_3+1+n}}, \quad n = 0, 1, \ldots$$

choose as test function in (5.2)

$$\varphi = -\left[(v_\omega)_h - k_n\right]_- \zeta_n^2$$

and proceed as in the proof of Lemma 5.3 to get

$$\text{ess sup} \int_{K_{R_n} \times \{t\}} (v_\omega - k_n)_-^2 \zeta_n^2 + \phi\left(\frac{\omega}{2^{s_3+2}}\right) \int_{-j_0 \frac{R_n^2}{\mu}}^{-(j_0-1)\frac{R_n^2}{\mu}} \int_{K_{R_n}} |\zeta_n \nabla(v_\omega - k_n)_-|^2$$

$$\leq C \frac{2^{2(n+1)}}{R^2} \left(\frac{\omega}{2^{s_3+2}}\right)^2 \phi\left(\frac{\omega}{2^{s_3}}\right) \int_{-j_0 \frac{R_n^2}{\mu}}^{-(j_0-1)\frac{R_n^2}{\mu}} \int_{K_{R_n}} \chi_{\{v_\omega \leq k_n\}}.$$

Now perform a change in the time variable, putting $\bar{t} = \left\{t + (j_0 - 1)\frac{R_n^2}{\mu}\right\}\mu$ and define

$$\overline{v_\omega}(\cdot, \bar{t}) = v_\omega(\cdot, t) \qquad \text{and} \qquad \overline{\zeta_n}(\cdot, \bar{t}) = \zeta_n(\cdot, t),$$

and obtain the inequality, with $\Gamma = \frac{\phi_1}{\phi_2}$,

$$\Gamma^{\frac{d+2}{2}} \text{ess sup} \int_{K_{R_n} \times \{t\}} (\overline{v_\omega} - k_n)_-^2 \overline{\zeta_n}^2 + \int_{-R_n^2}^0 \int_{K_{R_n}} |\overline{\zeta_n} \nabla(\overline{v_\omega} - k_n)_-|^2$$

$$\leq C \frac{2^{2(n+1)}}{R^2} \left(\frac{\omega}{2^{s_3}}\right)^2 \Gamma \int_{-R_n^2}^0 \int_{K_{R_n}} \chi_{\{\overline{v_\omega} \leq k_n\}}.$$

The conclusion of the proof follows from a refinement of the iteration technique used in the previous results; it can be found in [13]. ☐

Remark 5.14. Observe that (5.27) plays the same role as the assumption in the first alternative and that Lemma 5.13 is the corresponding analogue of Lemma 5.3. The next two results, leading to the conclusion of the second alternative, reproduce the ideas in Lemma 5.4 and Proposition 5.5, *i.e.*, we first use the logarithmic estimates to extend the result to a full cylinder in time and then, with the aid of the energy estimates, conclude that v is strictly away from 0 in a cylinder of the type $Q(\tau, \rho)$.

Lemma 5.15. *Given $\nu_1 \in (0,1)$, there exists $s_6 \in \mathbb{N}$, depending only on the data, such that*

$$\left| \left\{ x \in K_{\frac{R}{4}} : v(x,t) \le \frac{\omega}{2^{s_6}} \right\} \right| \le \nu_1 \left| K_{\frac{R}{4}} \right|, \quad \forall t \in \left(-j_0 \frac{\left(\frac{R}{2}\right)^2}{\mu}, 0 \right).$$

Proof. To simplify, put

$$\tilde{t} = j_0 \frac{\left(\frac{R}{2}\right)^2}{\mu}.$$

We use a logarithmic estimate for the function $(v-k)_-$ in the cylinder $Q(\tilde{t}, \frac{R}{2})$, with the choices

$$k = \frac{\omega}{2^{s_3+1}} \quad \text{and} \quad c = \frac{\omega}{2^{n+1}},$$

where $n \in \mathbb{N}$ will be chosen later, as parameters in the standard logarithmic function(see (2.7)).

Choose a piecewise smooth cutoff function $0 < \zeta(x) \le 1$, defined on $K_{\frac{R}{2}}$ and such that

$$\zeta = 1 \text{ in } K_{\frac{R}{4}} \quad \text{and} \quad |\nabla \zeta| \le \frac{C}{R},$$

and multiply (5.2) by $2\psi^-(v_h)(\psi^-)'(v_h)\zeta^2$. As in the proof of Lemma 5.4, with the obvious changes, we get

$$\sup_{-\tilde{t} \le t \le 0} \int_{K_{\frac{R}{2}} \times \{t\}} \left[\psi^-(v) \right]^2 \zeta^2 \le C(n - s_3) \phi\left(\frac{\omega}{2^{s_3+1}}\right) \frac{\tilde{t}}{R^2} \left| K_{\frac{R}{2}} \right|$$

$$\le C(n - s_3) \frac{\phi\left(\frac{\omega}{2^{s_3+1}}\right)}{\phi\left(\frac{\omega}{2^m}\right)} \left| K_{\frac{R}{2}} \right|$$

since, recalling (5.26),

$$\tilde{t} = j_0 \frac{\left(\frac{R}{2}\right)^2}{\mu} \le \frac{n_0}{\mu} \left(\frac{R}{2}\right)^2 = \frac{\left(\frac{R}{2}\right)^2}{\phi\left(\frac{\omega}{2^m}\right)}.$$

Now, since the integrand is nonnegative, we estimate below the left hand side of the inequality integrating over the smaller set

$$S = \left\{ x \in K_{\frac{R}{4}} : v(x,t) \le \frac{\omega}{2^{n+1}} \right\} \subset K_{\frac{R}{2}}$$

and, observing that in S, $\zeta = 1$ and

$$\left[\psi^-(v)\right]^2 \geq (n - s_3)^2 (\ln 2)^2,$$

we get

$$\left|\left\{x \in K_{\frac{R}{4}} : v(x,t) \leq \frac{\omega}{2^{n+1}}\right\}\right| \leq C \, \frac{1}{n - s_3} \, \frac{\phi\left(\frac{\omega}{2^{s_3+1}}\right)}{\phi\left(\frac{\omega}{2^m}\right)} \, \left|K_{\frac{R}{4}}\right|.$$

To prove the lemma choose

$$s_6 = n + 1 \qquad \text{with} \qquad n \geq s_3 + \frac{C}{\nu_1} \, \frac{\phi\left(\frac{\omega}{2^{s_3+1}}\right)}{\phi\left(\frac{\omega}{2^m}\right)}.$$

Concerning the independence of ω, observe that s_3 was chosen independently of ω and

$$\frac{\phi\left(\frac{\omega}{2^{s_3+1}}\right)}{\phi\left(\frac{\omega}{2^m}\right)} = \frac{\left(\frac{\omega}{2^{s_3+1}}\right)^{p_2}}{\left(\frac{\omega}{2^m}\right)^{p_2}} = 2^{(m-s_3-1)p_2},$$

expression that does not depended on ω.   □

Proposition 5.16. *There exists a constant $s_6 \in \mathbb{N}$, depending only on the data, such that*

$$v(x,t) > \frac{\omega}{2^{s_6}}, \quad \text{a.e. in} \quad Q\left(\tilde{t}, \frac{R}{8}\right).$$

Proof. Define

$$R_n = \frac{R}{8} + \frac{R}{2^{n+3}}, \quad n = 0,1,\dots$$

and construct the family of nested and shrinking cylinders $Q(\tilde{t}, R_n)$. Take piecewise smooth cutoff functions $0 < \zeta_n(x) \leq 1$, independent of t, defined in K_{R_n} and satisfying the assumptions

$$\zeta_n = 1 \text{ in } K_{R_{n+1}}, \quad |\nabla \zeta_n| \leq \frac{2^{n+4}}{R} \quad \text{and} \quad |\Delta \zeta_n| \leq \frac{2^{2(n+4)}}{R^2}.$$

Take also

$$k_n = \frac{\omega}{2^{s_6+1}} + \frac{\omega}{2^{s_6+1+n}}, \quad n = 0,1,\dots,$$

(where $s_6 > 1$ is to be chosen) and derive local energy inequalities similar to those obtained in Proposition 5.5, now for the functions

$$-(v_\omega - k_n)_- \zeta_n^2, \quad \text{with} \quad v_\omega := \max\left(v, \frac{\omega}{2^{s_6+1}}\right)$$

and in the cylinders $Q\left(\bar{t}, R_n\right)$. Using the same reasoning, we get

$$\sup_{-\tilde{t}<t<0} \int_{K_{R_n}\times\{t\}} (v_\omega - k_n)_+^2 \zeta_n^2 + \phi\left(\frac{\omega}{2^{s_6}}\right) \int_{-\tilde{t}}^0 \int_{K_{R_n}} |\zeta_n \nabla(v_\omega - k_n)_+|^2$$

$$\leq C\, \phi\left(\frac{\omega}{2^{s_6}}\right) \frac{2^{2(n+4)}}{R^2} \left(\frac{\omega}{2^{s_6+1}}\right)^2 \int_{-\tilde{t}}^0 \int_{K_{R_n}} \chi_{\{v_\omega \leq k_n\}}.$$

The change in the time variable $\bar{t} = \left(\dfrac{R}{2}\right)^2 \dfrac{t}{\tilde{t}}$, with the new function

$$\overline{v_\omega}(\cdot, \bar{t}) = v_\omega(\cdot, t),$$

leads to

$$\frac{\left(\frac{R}{2}\right)^2}{\phi\left(\frac{\omega}{2^{s_6}}\right)\tilde{t}} \sup_{-\left(\frac{R}{2}\right)^2<t<0} \int_{K_{R_n}\times\{t\}} (\overline{v_\omega} - k_n)_-^2 \zeta_n^2$$

$$+ \int_{-\left(\frac{R}{2}\right)^2}^0 \int_{K_{R_n}} |\zeta_n \nabla(\overline{v_\omega} - k_n)_-|^2$$

$$\leq C\, \frac{2^{2(n+4)}}{R^2} \left(\frac{\omega}{2^{s_6+1}}\right)^2 \int_{-\left(\frac{R}{2}\right)^2}^0 \int_{K_{R_n}} \chi_{\{\overline{v_\omega}\leq k_n\}},$$

after multiplication of the whole expression by $\dfrac{\left(\frac{R}{2}\right)^2}{\phi\left(\frac{\omega}{2^{s_6}}\right)\tilde{t}}$. Now, if s_6 is chosen

sufficiently large, we have

$$\frac{\left(\frac{R}{2}\right)^2}{\phi\left(\frac{\omega}{2^{s_6}}\right)\tilde{t}} \geq 1$$

since

$$\frac{\left(\frac{R}{2}\right)^2}{\phi\left(\frac{\omega}{2^{s_6}}\right)\tilde{t}} = \frac{\mu}{j_0\,\phi\left(\frac{\omega}{2^{s_6}}\right)} \geq \frac{\mu}{n_0\,\phi\left(\frac{\omega}{2^{s_6}}\right)} = \frac{\phi\left(\frac{\omega}{2^m}\right)}{\phi\left(\frac{\omega}{2^{s_6}}\right)},$$

which shows also that it suffices to pick $s_6 > m$.

The rest of the proof follows the same lines of the proof of Proposition 5.5, that consist basically in using the iteration technique and then Lemma 5.15. We omit the details. □

Corollary 5.17. *There exists a constant $\sigma_1 \in (0,1)$, depending only on the data, such that*

$$\operatorname*{ess\,osc}_{Q\left(\bar{t}, \frac{R}{8}\right)} v \leq \sigma_1\, \omega.$$

Proof. It is similar to the proof of Corollary 5.7. We find $\sigma_1 = \left(1 - \frac{1}{2^{s_6}}\right)$. □

An immediate consequence of Corollaries 5.7 and 5.17 is our final result.

Proposition 5.18. *There exists a constant $\sigma \in (0,1)$, that depends only on the data, and a cylinder $Q\left(t^\diamond, \frac{R}{8}\right)$, such that*

$$\operatorname*{ess\,osc}_{Q(t^\diamond, \frac{R}{8})} v \leq \sigma\,\omega.$$

Proof. Since one of (5.7) or (5.20) has to be true, the conclusion of at least one of Corollaries 5.7 or 5.17 holds. Choosing

$$\sigma = \max\left\{\sigma_0\,, \sigma_1\right\},$$

and

$$t^\diamond = \min\left\{\widehat{t}, \widetilde{t}\right\},$$

we obtain the conclusion. □

The proof of Theorem 5.2 now follows from Proposition 5.18 as in Section 4.4. We stress that the Hölder continuity is obtained since σ in Proposition 5.18 is independent of the oscillation ω.

5.5 A Problem in Chemotaxis

We now consider a similar problem

$$\begin{cases} v_t - \operatorname{div}\left(\alpha(v)\nabla v\right) = -\operatorname{div}\left(\chi v f(v)\nabla u\right) \\[2mm] u_t - \Delta u = g(v, u) \end{cases} \quad \text{in} \quad \Omega_T \quad (5.28)$$

arising from the modelling of chemotaxis, a property of certain living organisms to be repelled or attracted to chemical substances. Here $v = v(x,t)$ represents the density of a cell-population, $u = u(x,t)$ represents the chemoattractant (repellent) concentration, $\alpha(v)$ is a density-dependent diffusion coefficient and $f(v(x,t))$ measures the probability that a cell in position x at time t finds space in its neighboring location. The cells are attracted by the chemical and χ denotes their chemotactic sensitivity. The function $g(v, u)$ describes the rates of production and degradation of the chemoattractant.

The above model is a special case of the original Keller-Segel model [32], introduced to describe the aggregation of the cellular slime mold *Dictyostelium discoideum* due to the effect of the cyclic Adenosine Monophosphate (cAMP), an attractive chemical signal for the amoebae. We consider the following assumptions, corresponding to the two the main features of the model:

- There is a maximal density of cells, the threshold v_m, such that $f(v_m) = 0$. Intuitively, this amounts to a switch to repulsion at high densities, sometimes referred to as volume–filling effect or prevention of overcrowding (see [29]). This threshold condition has a clear biological interpretation:

the cells stop to accumulate at a given point of Ω after their density attains certain threshold values and the chemotactic cross diffusion $h(v) = \chi v f(v)$ vanishes identically when $v \geq v_m$.

- The density-dependent diffusion coefficient $\alpha(v)$ degenerates for $v = 0$ and $v = v_m$. This means, in particular, that there is no diffusion when v approaches values close to the threshold.

This interpretation was proposed in [6], where the diffusion coefficient takes the form $\alpha(v) = \epsilon v(1 - v)$, for $\epsilon > 0$. The main advantages of the nonlinear diffusion model seem to be related to the finite speed of propagation (which is more realistic in biological applications than infinite speed) and the asymptotic behavior of solutions.

Defining new variables through

$$\tilde{v} = \frac{v}{v_m}, \quad \tilde{u} = u,$$

we have $\tilde{v}_m = 1$. After performing this linear transformation, we omit the tildes in the notation. A typical example of f in this case is $f(v) = 1 - v$.

Under suitable assumptions on the data, an existence theorem is proved in [4], using a Schauder fixed-point argument on a regularized problem and the compactness method. Here, we comment on the local Hölder continuity of a weak solution v of the system under the same assumptions (A1)–(A3) on α used for the immiscible fluids model. The novelty with respect to the previous sections is the additional lower-order term $\operatorname{div}(\chi v f(v) \nabla u)$; we show that it satisfies the appropriate growth conditions due to its special form and the available regularity for u, which follows from the classical theory of parabolic PDEs. The pertinent definition of local weak solution can be cast in the following formulation:
for every compact $K \subset \Omega$ and for every $0 < t < T - h$,

$$\int_{K \times \{t\}} \{(v_h)_t \, \varphi + (\alpha(v) \nabla v)_h \cdot \nabla \varphi - \chi \, (v f(v))_h \, \nabla u \cdot \nabla \varphi\} \, dx = 0, \quad (5.29)$$

for all $\varphi \in H_0^1(K)$.

Here, u is treated as a given function in its existence class so all the terms in the above expression have a meaning. To study the locally regularity of v we proceed exactly as before, defining the same intrinsic geometry and establishing the same type of alternative. To illustrate this fact, we now prove the equivalent of Lemma 5.3.

Recall that we denote

$$v_\omega := \min\left\{v, 1 - \frac{\omega}{4}\right\}.$$

Taking the cylinder for which the first alternative (5.7) holds, we define

$$R_n = \frac{R}{2} + \frac{R}{2^{n+1}}, \quad n = 0, 1, \ldots,$$

and construct the family of nested and shrinking cylinders

$$Q_{R_n}^{t^*} = K_{R_n} \times \left(t^* - \frac{R_n^2}{\psi(\frac{\omega}{4})}, t^*\right).$$

Next, we consider piecewise smooth cutoff functions $0 \leq \xi_n \leq 1$, defined in these cylinders, and satisfying assumptions (5.8). Letting

$$k_n = 1 - \frac{\omega}{4} - \frac{\omega}{2^{n+2}}, \quad n = 0, 1, \ldots$$

we choose as test function in (5.29) $\varphi = [(v_\omega)_h - k_n]_+ \xi_n^2$ and integrate in time over $(t^* - \frac{R_n^2}{\psi(\frac{\omega}{4})}, t)$ for $t \in (t^* - \frac{R_n^2}{\psi(\frac{\omega}{4})}, t^*)$ with $K = K_{R_n}$. To simplify the notation, we put

$$\tau_n := t^* - \frac{R_n^2}{\psi(\frac{\omega}{4})},$$

and again omit, from here on, dx and dt in all integrals.

We only consider the lower order term since it encompasses the main novelty with respect to the previous sections. After passing to the limit in h, using the convergence properties of the Steklov average and Young's inequality, we obtain

$$\chi \int_{\tau_n}^t \int_{K_{R_n}} vf(v)\nabla u \cdot \left\{\xi_n^2 \nabla (v_\omega - k_n)_+ + 2(v_\omega - k_n)_+ \xi_n \nabla \xi_n\right\}$$

$$\leq \frac{1}{2}\psi\left(\frac{\omega}{4}\right) \int_{\tau_n}^t \int_{K_{R_n}} |\xi_n \nabla (v_\omega - k_n)_+|^2 + \frac{1}{2\psi\left(\frac{\omega}{4}\right)} M^2 \int_{\tau_n}^t \int_{K_{R_n}} |\nabla u|^2 \chi_{\{v_\omega \geq k_n\}}$$

$$+ 2M \int_{\tau_n}^t \int_{K_{R_n}} |\nabla u||\nabla \xi_n| \left(\frac{\omega}{4}\right) \chi_{\{v_\omega \geq k_n\}}$$

since $(v_\omega - k_n)_+ \leq \frac{\omega}{4}$, and defining $M := \|\chi vf(v)\|_{L^\infty(\Omega_T)}$. Using again Young's inequality, we bound from above by

$$\frac{1}{2}\psi\left(\frac{\omega}{4}\right) \int_{\tau_n}^t \int_{K_{R_n}} |\xi_n \nabla (v_\omega - k_n)_+|^2 + \frac{M^2 + 2M}{2\psi\left(\frac{\omega}{4}\right)} \int_{\tau_n}^t \int_{K_{R_n}} |\nabla u|^2 \chi_{\{v_\omega \geq k_n\}}$$

$$+ M\frac{2^{2(n+1)}}{R^2} \left(\frac{\omega}{4}\right)^2 \psi\left(\frac{\omega}{4}\right) \int_{\tau_n}^t \int_{K_{R_n}} \chi_{\{v_\omega \geq k_n\}}.$$

We conclude with the estimate

$$\int_{\tau_n}^{t} \int_{K_{R_n}} |\nabla u|^2 \chi_{\{v_\omega \geq k_n\}} \leq \|\nabla u\|_{L^p(\Omega_T)}^2 \left\{ \int_{\tau_n}^{t} \left| A_{k_n, R_n}^+(\sigma) \right| d\sigma \right\}^{1-\frac{2}{p}}$$

where

$$A_{k_n, R_n}^+(\sigma) := \{x \in K_{R_n} : v(x, \sigma) > k_n\}$$

and p is chosen sufficiently large. This is possible since, from standard parabolic theory,

$$u \in L^q\left(0, T; W^{2,q}(\Omega)\right), \qquad \text{for all } q > 1.$$

Putting these estimate together with the ones obtained in the previous section, we arrive at

$$\operatorname*{ess\,sup}_{\tau_n \leq t \leq t^*} \int_{K_{R_n} \times \{t\}} (v_\omega - k_n)_+^2 \, \xi_n^2 + \psi\left(\frac{\omega}{4}\right) \int_{\tau_n}^{t^*} \int_{K_{R_n}} \left| \xi_n \nabla (v_\omega - k_n)_+ \right|^2$$

$$\leq 2 \left\{ \frac{[C_1 \psi(\frac{\omega}{2})]^2}{(C_0 - 1)\psi(\frac{\omega}{4})} + (7 + M)\psi\left(\frac{\omega}{4}\right) \right\} \frac{2^{2(n+1)}}{R^2} \left(\frac{\omega}{4}\right)^2 \int_{\tau_n}^{t^*} \int_{K_{R_n}} \chi_{\{v_\omega \geq k_n\}}$$

$$+ \frac{M^2 + 2M}{\psi\left(\frac{\omega}{4}\right)} \|\nabla u\|_{L^p(\Omega_T)}^2 \left\{ \int_{\tau_n}^{t^*} \left| A_{k_n, R_n}^+(\sigma) \right| d\sigma \right\}^{1-\frac{2}{p}}.$$

Next we perform a change in the time variable, putting $\bar{t} = (t - t^*)\psi(\frac{\omega}{4})$, and define

$$\overline{v_\omega}(\cdot, \bar{t}) = v_\omega(\cdot, t) \qquad \text{and} \qquad \overline{\xi_n}(\cdot, \bar{t}) = \xi_n(\cdot, t),$$

to obtain the simplified inequality

$$\left\| (\overline{v_\omega} - k_n)_+ \, \overline{\xi_n} \right\|_{V^2(Q(R_n^2, R_n))}^2$$

$$\leq 2 \left\{ \frac{C_1^2}{(C_0 - 1)} \left[\frac{\psi(\frac{\omega}{2})}{\psi(\frac{\omega}{4})} \right]^2 + 7 + M \right\} \frac{2^{2(n+1)}}{R^2} \left(\frac{\omega}{4}\right)^2 A_n$$

$$+ (M^2 + 2M)\|\nabla u\|_{L^p(\Omega_T)}^2 \left[\psi\left(\frac{\omega}{4}\right) \right]^{\frac{2}{p} - 2} A_n^{1 - \frac{2}{p}},$$

defining, for each n,

$$A_n = \int_{-R_n^2}^{0} \int_{K_{R_n}} \chi_{\{\overline{v_\omega} \geq k_n\}} dx \, d\bar{t}.$$

Next, observe that the following estimates hold

$$\frac{1}{2^{2(n+2)}}\left(\frac{\omega}{4}\right)^2 A_{n+1} = |k_{n+1} - k_n|^2\, A_{n+1} \tag{5.30}$$

$$\leq \left\|(\overline{v_\omega} - k_n)_+\right\|^2_{2,Q(R^2_{n+1},R_{n+1})}$$

$$\leq \left\|(\overline{v_\omega} - k_n)_+ \overline{\xi_n}\right\|^2_{2,Q(R^2_n,R_n)}$$

$$\leq C\, \left\|(\overline{v_\omega} - k_n)_+ \overline{\xi_n}\right\|^2_{V^2(Q(R^2_n,R_n))}\, A_n^{\frac{2}{d+2}}$$

$$\leq 2C\left\{\frac{C_1^2}{(C_0-1)}\left[\frac{\psi(\frac{\omega}{2})}{\psi(\frac{\omega}{4})}\right]^2 + 7 + M\right\}\frac{2^{2(n+1)}}{R^2}\left(\frac{\omega}{4}\right)^2 A_n^{1+\frac{2}{d+2}}$$

$$+C(M^2+2M)\|\nabla u\|^2_{L^p(\Omega_T)}\left[\psi\left(\frac{\omega}{4}\right)\right]^{\frac{2}{p}-2} A_n^{1-\frac{2}{p}+\frac{2}{d+2}}.$$

The reasoning is the same as before. Next, define the numbers

$$X_n = \frac{A_n}{|Q(R^2_n,R_n)|} \quad ; \quad Z_n = \frac{A_n^{1/p}}{|K_{R_n}|},$$

divide (5.30) by $\left|Q(R^2_{n+1},R_{n+1})\right|$ and obtain the recursive relation

$$X_{n+1} \leq \gamma\, 4^{2n}\left\{X_n^{1+\frac{2}{d+2}} + X_n^{\frac{2}{d+2}} Z_n^{1+\kappa}\right\}, \quad n = 0,1,2,\dots$$

where $\kappa = p - 3 > 0$ and

$$\gamma = C\,\max\left\{\frac{C_1^2}{(C_0-1)}\left[\frac{\psi(\frac{\omega}{2})}{\psi(\frac{\omega}{4})}\right]^2 + 7 + M\; ; \; \left(\frac{\omega}{4}\right)^{-2}\left[\psi\left(\frac{\omega}{4}\right)\right]^{\frac{2}{p}-2} R^{d\kappa}\right\}.$$

A similar reasoning leads to

$$Z_{n+1} \leq \gamma\, 4^{2n}\left\{X_n + Z_n^{1+\kappa}\right\}, \quad n = 0,1,2,\dots$$

We can now use Lemma 2.10 to conclude that if

$$X_0 + Z_0^{1+\kappa} \leq (2\gamma)^{-\frac{1+\kappa}{\theta}} 4^{\frac{-2(1+\kappa)}{\theta^2}} =: \nu_0, \qquad \theta = \min\left\{\frac{2}{d+2}; \kappa\right\} \tag{5.31}$$

then

$$X_n, Z_n \longrightarrow 0. \tag{5.32}$$

But (5.31) follows from the assumption in the first alternative and the conclusion is a direct consequence of (5.32).

It remains to show that ν_0, $i.e.$, γ, is independent of ω, which is crucially related to the fact that ψ is a power. In fact, we have

$$\frac{\psi(\frac{\omega}{2})}{\psi(\frac{\omega}{4})} = \left(\frac{\omega}{2}\right)^{p_1}\left(\frac{4}{\omega}\right)^{p_1} = 2^{p_1}.$$

On the other hand, we can assume, without loss of generality, that

$$\left(\frac{\omega}{4}\right)^{-2}\left[\psi\left(\frac{\omega}{4}\right)\right]^{\frac{2}{p}-2} R^{d\kappa} \leq 1.$$

Otherwise, we would have $\omega < CR^{\alpha}$, with $\alpha = \frac{d\kappa p}{2p+2pp_1-2p_1} > 0$, and the result would be trivial.

6

Flows in Porous Media: The Variable Exponent Case

Partial differential equations with nonlinearities involving variable exponents have attracted an increasing amount of attention in recent years. The development, mainly by Růžička [47], of a theory modeling the behavior of electrorheological fluids, an important class of non-Newtonian fluids, seems to have boosted a still far from completed effort to study and understand this type of equations. Other important applications relate to image processing [8], elasticity [56] or flows in porous media [2].

We will consider the parabolic equation in divergence form

$$u_t - \operatorname{div}\left(|u|^{\gamma(x,t)}\nabla u\right) = 0, \tag{6.1}$$

with a variable exponent of nonlinearity γ, which is a generalization of the famous porous medium equation and occurs as a model for the flow of an ideal barotropic gas through a porous medium. The main feature in equation (6.1) is that it is degenerate due to the exponential nonlinearity: the diffusion coefficient $|u|^{\gamma(x,t)}$ vanishes at points where $u = 0$. Results on the existence and uniqueness of weak solutions of (6.1), together with some localization properties, were obtained by Antontsev and Shmarev [2]. Under appropriate assumptions, we use intrinsic scaling to prove that weak solutions are locally continuous.

6.1 The Porous Medium Equation in its Own Geometry

We assume the exponent γ satisfies the following assumptions:

(A1) $\gamma \in L^\infty\left(0, T; W^{1,p}(\Omega)\right)$, for some $p > \max\{2, d\}$;

(A2) For constants $\gamma^-, \gamma^+ > 0$,

$$0 < \gamma^- \leq \gamma(x,t) \leq \gamma^+ < \infty, \quad \text{a.e. } (x,t) \in \Omega_T.$$

It is proved in [2], under less restrictive assumptions on γ, that there exists a unique solution to the initial boundary value problem associated with (6.1) and that the solution is bounded. It is also shown that the solution is nonnegative if the initial data is nonnegative and that is why it is reasonable to assume $u(x,t) \in [0,1]$ a.e. in Ω_T. Here we consider local weak solutions.

Definition 6.1. *A measurable function u is a local weak solution of (6.1) if*

(i) $u \in L^\infty \left(0,T; L^\infty(\Omega)\right)$ *with* $u(x,t) \in [0,1]$ *a.e. in Ω_T;*

(ii) $u \in C\left(0,T; L^2(\Omega)\right)$ *and* $u^{\frac{\gamma(x,t)}{2}} \nabla u \in L^2\left(0,T; L^2(\Omega)\right)$;

(iii) *for every compact $K \subset \Omega$ and for every subinterval $[t_1, t_2] \subset (0,T]$,*

$$\int_K u\phi \, dx \Big|_{t_1}^{t_2} + \int_{t_1}^{t_2} \int_K \left\{ -u\phi_t + u^{\gamma(x,t)} \nabla u \cdot \nabla \phi \right\} dx \, dt = 0, \qquad (6.2)$$

for all $\phi \in H^1_{\text{loc}}\left(0,T; L^2(K)\right) \cap L^2_{\text{loc}}\left(0,T; H^1_0(K)\right)$.

As before we use the Steklov average to obtain the following formulation which is equivalent to (iii):

(iii)' *for every compact $K \subset \Omega$ and for every $0 < t < T - h$,*

$$\int_{K \times \{t\}} \left\{ (u_h)_t \, \phi + \left(u^{\gamma(x,\cdot)} \nabla u \right)_h \cdot \nabla \phi \right\} dx = 0, \qquad (6.3)$$

for all $\phi \in H^1_0(K)$.

We start by defining an intrinsic geometric configuration tailored for this specific PDE. Let (x_0, t_0) be a point of the space-time domain Ω_T that, by translation, we may assume to be $(0,0)$. Consider small positive numbers $\epsilon > 0$ and $R > 0$ such that the cylinder $Q\left(R^{2-\epsilon}, R\right) \subset \Omega_T$ and define

$$\mu^- := \operatorname*{ess\,inf}_{Q(R^{2-\epsilon},R)} u \; ; \quad \mu^+ := \operatorname*{ess\,sup}_{Q(R^{2-\epsilon},R)} u \; ; \quad \omega := \operatorname*{ess\,osc}_{Q(R^{2-\epsilon},R)} u = \mu^+ - \mu^-.$$

Recalling that (6.1) is degenerate at the points where $u = 0$, the interesting case to investigate is when $\mu^- = 0$ and, consequently, $\mu^+ = \omega$. From now on, we will assume this is in force. Construct the cylinder

$$Q(a_0 R^2, R), \qquad a_0 = \left(\frac{4}{\omega}\right)^{\gamma^+},$$

and assume that

$$\omega \geq 4R^{\frac{\epsilon}{\gamma^+}}. \qquad (6.4)$$

This implies that $Q(a_0 R^2, R) \subset Q\left(R^{2-\epsilon}, R\right)$ and then

$$\operatorname*{ess\,osc}_{Q\left(a_0 R^2, R\right)} u \leq \omega. \tag{6.5}$$

Remark 6.2. If (6.4) does not hold, then the oscillation ω goes to zero when the radius R goes to zero, in a way given by the reverse inequality, and there is nothing to prove. When $\gamma \equiv 0$, $a_0 = 1$ and we recover the standard parabolic cylinder with the natural homogeneity of the space and time variables.

Given $\nu_0 \in (0, 1)$, to be determined in terms of the data and ω, either

$$\left|(x, t) \in Q(a_0 R^2, R) : u(x, t) < \frac{\omega}{2}\right| \leq \nu_0 \left|Q(a_0 R^2, R)\right| \tag{6.6}$$

or, noting that $\mu^+ - \frac{\omega}{2} = \frac{\omega}{2}$,

$$\left|(x, t) \in Q(a_0 R^2, R) : u(x, t) > \mu^+ - \frac{\omega}{2}\right| < (1 - \nu_0) \left|Q(a_0 R^2, R)\right|. \tag{6.7}$$

The analysis of this alternative leads to the following result.

Proposition 6.3. *There exist positive numbers $\nu_0, \sigma \in (0, 1)$, depending on the data and on ω, such that*

$$\operatorname*{ess\,osc}_{Q\left(\frac{\nu_0}{2} a_0 \left(\frac{R}{2}\right)^2, \frac{R}{2}\right)} u \leq \sigma \omega. \tag{6.8}$$

An immediate consequence is the following.

Theorem 6.4. *Under assumptions $(A1) - (A2)$ any locally bounded weak solution of (6.1) is locally continuous in Ω_T.*

The proof follows from a slight modification of the arguments in Section 4.4. From (6.8) one defines recursively a sequence Q_n of nested and shrinking cylinders and a sequence ω_n converging to zero (see also the proof of Theorem 7.7 on Chapter 7), such that

$$\operatorname*{ess\,osc}_{Q_n} u \leq \omega_n.$$

This is enough to obtain the continuity of u but we are unable to derive a modulus since the constant σ appearing in Proposition 6.3 depends on the oscillation ω.

6.2 Reducing the Oscillation

Assume that (6.6) is verified. In the following, we determine the number ν_0 and guarantee that the solution u is above a smaller level within a smaller cylinder.

Proposition 6.5. *There exists $\nu_0 \in (0,1)$, depending only on the data and ω, such that if (6.6) holds then*

$$u(x,t) > \frac{\omega}{4}, \quad a.e. \ (x,t) \in Q\left(a_0 \left(\frac{R}{2}\right)^2, \frac{R}{2}\right). \tag{6.9}$$

Proof. This proof, as well as the subsequent ones, will be presented for the particular case $p = \infty$ in assumption (A1). The case of a general $p > d$ is similar but technically more involved (see Remark 6.7).

Define two decreasing sequences of positive numbers

$$R_n = \frac{R}{2} + \frac{R}{2^{n+1}}, \quad k_n = \frac{\omega}{4} + \frac{\omega}{2^{n+2}}, \quad n = 0,1,\ldots$$

and construct the family of nested and shrinking cylinders $Q_n = Q(a_0 R_n^2, R_n)$. Introduce the function $u_\omega = \max\{u, \frac{\omega}{4}\}$. In the weak formulation (6.3) take

$$\phi = -\left((u_\omega)_h - k_n\right)_- \xi_n^2,$$

where $0 \le \xi_n \le 1$ are smooth cutoff functions defined in Q_n and satisfying

$$\begin{cases} \xi_n \equiv 1 \quad \text{in} \quad Q_{n+1}, \qquad \xi_n \equiv 0 \quad \text{on the parabolic boundary of } Q_n \\[2mm] |\nabla \xi_n| \le \frac{2^{n+2}}{R}, \qquad |\Delta \xi_n| \le \frac{2^{2(n+2)}}{R^2}, \qquad 0 < (\xi_n)_t \le \frac{2^{2(n+2)}}{a_0 R^2}, \end{cases}$$

and integrate in time over $(-a_0 R_n^2, t)$, for $t \in (-a_0 R_n^2, 0)$. We obtain (omitting the dx and dt in all integrals from now on)

$$I_1 + I_2 := \int_{-a_0 R_n^2}^{t} \int_{K_{R_n}} (u_h)_t \left[-\left((u_\omega)_h - k_n\right)_- \xi_n^2\right]$$

$$+ \int_{-a_0 R_n^2}^{t} \int_{K_{R_n}} (u^\gamma \nabla u)_h \cdot \nabla \left[-\left((u_\omega)_h - k_n\right)_- \xi_n^2\right] = 0.$$

Concerning the first integral, we have

$$I_1 = \int_{-a_0 R_n^2}^{t} \int_{K_{R_n}} (u_h)_t \left[-\left((u_\omega)_h - k_n\right)_- \xi_n^2\right] \chi_{[(u_\omega)_h = u_h]}$$

$$+ \int_{-a_0 R_n^2}^{t} \int_{K_{R_n}} (u_h)_t \left[-\left((u_\omega)_h - k_n\right)_- \xi_n^2\right] \chi_{[(u_\omega)_h = \frac{\omega}{4}]}$$

$$= \frac{1}{2} \int_{-a_0 R_n^2}^{t} \int_{K_{R_n}} \left(\left((u_\omega)_h - k_n\right)_-^2\right)_t \xi_n^2$$

$$+ \left(\frac{\omega}{2^{n+2}}\right) \int_{-a_0 R_n^2}^{t} \int_{K_{R_n}} \left(\left(u_h - \frac{\omega}{4}\right)_-\right)_t \xi_n^2.$$

Next, we integrate by parts and let $h \to 0$. Using Lemma 2.2, we get

$$\frac{1}{2} \int_{K_{R_n} \times \{t\}} (u_\omega - k_n)^2_- \xi_n^2 - \int_{-a_0 R_n^2}^t \int_{K_{R_n}} (u_\omega - k_n)^2_- \xi_n (\xi_n)_t$$

$$+ \left(\frac{\omega}{2^{n+2}}\right) \int_{K_{R_n} \times \{t\}} \left(u - \frac{\omega}{4}\right)_- \xi_n^2 - 2 \left(\frac{\omega}{2^{n+2}}\right) \int_{-a_0 R_n^2}^t \int_{K_{R_n}} \left(u - \frac{\omega}{4}\right)_- \xi_n (\xi_n)_t$$

$$\geq \frac{1}{2} \int_{K_{R_n} \times \{t\}} (u_\omega - k_n)^2_- \xi_n^2 - 3 \left(\frac{\omega}{4}\right)^2 \frac{2^{2(n+2)}}{a_0 R^2} \int_{-a_0 R_n^2}^t \int_{K_{R_n}} \chi_{[u_\omega \leq k_n]},$$

since the third term is nonnegative and, for $0 \leq u \leq \frac{\omega}{4}$, $u_\omega = \frac{\omega}{4} \leq k_n$ and, for $\frac{\omega}{4} < u = u_\omega \leq k_n$,

$$(u_\omega - k_n)_- \leq k_n - u_\omega = k_n - u < k_n - \frac{\omega}{4} = \frac{\omega}{2^{n+2}} \leq \frac{\omega}{4}.$$

Concerning I_2, we first pass to the limit in h to get

$$I_2 \to \int_{-a_0 R_n^2}^t \int_{K_{R_n}} u^\gamma \nabla u \cdot \nabla \left(-(u_\omega - k_n)_- \xi_n^2\right)$$

$$= \int_{-a_0 R_n^2}^t \int_{K_{R_n}} u^\gamma \nabla u \cdot \nabla \left(-(u_\omega - k_n)_- \xi_n^2\right) \chi_{[u_\omega = u]}$$

$$+ \int_{-a_0 R_n^2}^t \int_{K_{R_n}} u^\gamma \nabla u \cdot \nabla \left(-(u_\omega - k_n)_- \xi_n^2\right) \chi_{[u_\omega = \frac{\omega}{4}]}$$

$$= \int_{-a_0 R_n^2}^t \int_{K_{R_n}} u_\omega^\gamma |\nabla (u_\omega - k_n)_-|^2 \xi_n^2$$

$$+ 2 \int_{-a_0 R_n^2}^t \int_{K_{R_n}} u_\omega^\gamma \nabla (u_\omega - k_n)_- \cdot \nabla \xi_n \xi_n (u_\omega - k_n)_-$$

$$+ 2 \left(\frac{\omega}{2^{n+2}}\right) \int_{-a_0 R_n^2}^t \int_{K_{R_n}} -u^\gamma \nabla u \cdot \nabla \xi_n \xi_n \chi_{[u \leq \frac{\omega}{4}]}$$

$$\geq \frac{1}{2} \int_{-a_0 R_n^2}^t \int_{K_{R_n}} u_\omega^\gamma |\nabla (u_\omega - k_n)_-|^2 \xi_n^2$$

$$- 2 \int_{-a_0 R_n^2}^t \int_{K_{R_n}} u_\omega^\gamma |\nabla \xi_n|^2 (u_\omega - k_n)_-^2$$

$$+ 2 \left(\frac{\omega}{2^{n+2}}\right) \int_{-a_0 R_n^2}^t \int_{K_{R_n}} \nabla \left(\int_u^{\frac{\omega}{4}} s^\gamma \, ds\right) \cdot \nabla \xi_n \xi_n \chi_{[u \leq \frac{\omega}{4}]}$$

$$+ 2 \left(\frac{\omega}{2^{n+2}}\right) \int_{-a_0 R_n^2}^t \int_{K_{R_n}} \left(\int_u^{\frac{\omega}{4}} (-\ln s) s^\gamma \, ds\right) \nabla \gamma \cdot \nabla \xi_n \xi_n \chi_{[u \leq \frac{\omega}{4}]}$$

and this is bounded from below by

$$I_2' := \frac{1}{2} \int_{-a_0 R_n^2}^t \int_{K_{R_n}} u_\omega^\gamma |\nabla (u_\omega - k_n)_-|^2 \xi_n^2$$

$$-2 \int_{-a_0 R_n^2}^t \int_{K_{R_n}} u_\omega^\gamma |\nabla \xi_n|^2 (u_\omega - k_n)_-^2$$

$$-2 \left(\frac{\omega}{2^{n+2}} \right) \int_{-a_0 R_n^2}^t \int_{K_{R_n}} \left(\int_u^{\frac{\omega}{4}} s^\gamma \, ds \right) (\xi_n |\Delta \xi_n| + |\nabla \xi_n|^2) \, \chi_{[u \le \frac{\omega}{4}]}$$

$$- \left(\frac{\omega}{2^{n+2}} \right) \int_{-a_0 R_n^2}^t \int_{K_{R_n}} \left(\int_u^{\frac{\omega}{4}} (-\ln s) s^\gamma \, ds \right) |\nabla \gamma|^2 \xi_n \chi_{[u \le \frac{\omega}{4}]}$$

$$- \left(\frac{\omega}{2^{n+2}} \right) \int_{-a_0 R_n^2}^t \int_{K_{R_n}} \left(\int_u^{\frac{\omega}{4}} (-\ln s) s^\gamma \, ds \right) |\nabla \xi_n|^2 \xi_n \chi_{[u \le \frac{\omega}{4}]}.$$

The first inequality is obtained by means of Cauchy's inequality with $\epsilon = \frac{1}{4}$, and the observation that, within the set $[u \le \frac{\omega}{4}]$,

$$\nabla \left(\int_u^{\frac{\omega}{4}} s^\gamma \, ds \right) = -u^\gamma \nabla u - \left(\int_u^{\frac{\omega}{4}} (-\ln s) s^\gamma \, ds \right) \nabla \gamma.$$

The second inequality is a consequence of integration by parts and again Cauchy's inequality, this time with $\epsilon = \frac{1}{2}$.

Next, observe that for $\frac{\omega}{4} < u = u_\omega \le k_n$, we get

$$(u_\omega - k_n)_- = k_n - u_\omega = k_n - u < \frac{\omega}{2^{n+2}} \le \frac{\omega}{4}$$

and

$$\frac{1}{a_0} \le \left(\frac{\omega}{4} \right)^\gamma < u_\omega^\gamma \le 1 \qquad \text{using} \quad (A2).$$

For $u \le \frac{\omega}{4}$, we get

$$u_\omega = \frac{\omega}{4} \le k_n;$$

$$\int_u^{\frac{\omega}{4}} s^\gamma \, ds \le \left(\frac{\omega}{4} \right)^\gamma \left(\frac{\omega}{4} - u \right) \le \frac{\omega}{4};$$

$$\int_u^{\frac{\omega}{4}} (-\ln s) s^\gamma \, ds \le \left(\frac{\omega}{4} \right)^\gamma \int_u^{\frac{\omega}{4}} (-\ln s) \, ds \le \int_u^{\frac{\omega}{4}} (-\ln s) \, ds$$

$$= -\frac{\omega}{4} \ln \left(\frac{\omega}{4} \right) + u \ln u + \frac{\omega}{4} - u \le \left(\frac{\omega}{4} - u \right) \left\{ \ln \left(\frac{4}{\omega} \right) + 1 \right\} \le \left(\frac{\omega}{4} \right) \frac{4}{\omega} = 1.$$

Using these inequalities, recalling the conditions on ξ_n, and the fact that $\omega \le 1$, I_2' is bounded from below by

$$\frac{1}{2a_0} \int_{-a_0 R_n^2}^t \int_{K_{R_n}} |\nabla(u_\omega - k_n)_-|^2 \xi_n^2$$

$$-\frac{C(M)}{\omega} \left(\frac{\omega}{4}\right)^2 \frac{2^{2(n+2)}}{R^2} \int_{-a_0 R_n^2}^t \int_{K_{R_n}} \chi_{[u_\omega \le k_n]},$$

with $M = \|\gamma\|_{L^\infty(0,T;W^{1,\infty}(\Omega))}$. Combining the above results, we obtain the energy estimates

$$\sup_{-a_0 R_n^2 < t < 0} \int_{K_{R_n} \times \{t\}} (u_\omega - k_n)_-^2 \xi_n^2 + \frac{1}{a_0} \int_{-a_0 R_n^2}^0 \int_{K_{R_n}} |\nabla(u_\omega - k_n)_-|^2 \xi_n^2$$

$$\le 6 \left(\frac{\omega}{4}\right)^2 \frac{2^{2(n+2)}}{a_0 R^2} \int_{-a_0 R_n^2}^0 \int_{K_{R_n}} \chi_{[u_\omega \le k_n]}$$

$$+ \frac{C(M)}{\omega} \left(\frac{\omega}{4}\right)^2 \frac{2^{2(n+2)}}{R^2} \int_{-a_0 R_n^2}^0 \int_{K_{R_n}} \chi_{[u_\omega \le k_n]}.$$

Let us now consider the change of variables

$$z = \frac{t}{a_0}$$

and define the new functions

$$\bar{u}_\omega(x, z) = u_\omega(x, a_0 z) \quad ; \quad \bar{\xi}_n(x, z) = \xi_n(x, a_0 z).$$

Then the above estimates read

$$\sup_{-R_n^2 < z < 0} \int_{K_{R_n} \times \{z\}} (\bar{u}_\omega - k_n)_-^2 \bar{\xi}_n^2 + \int_{-R_n^2}^0 \int_{K_{R_n}} |\nabla(\bar{u}_\omega - k_n)_-|^2 \bar{\xi}_n^2$$

$$\le 6 \left(\frac{\omega}{4}\right)^2 \frac{2^{2(n+2)}}{R^2} \int_{-R_n^2}^0 \int_{K_{R_n}} \chi_{[\bar{u}_\omega \le k_n]}$$

$$+ \frac{C(M, \gamma^+)}{\omega^{1+\gamma^+}} \left(\frac{\omega}{4}\right)^2 \frac{2^{2(n+2)}}{R^2} \int_{-R_n^2}^0 \int_{K_{R_n}} \chi_{[\bar{u}_\omega \le k_n]}$$

$$\le \frac{C(M, \gamma^+)}{\omega^{1+\gamma^+}} \left(\frac{\omega}{4}\right)^2 \frac{2^{2(n+2)}}{R^2} A_n,$$

for $A_n := \int_{-R_n^2}^0 \int_{K_{R_n}} \chi_{[\bar{u}_\omega \le k_n]}$. These estimates imply the inequality

$$\|(\bar{u}_\omega - k_n)_-\|_{V^2(Q(R_{n+1}^2, R_{n+1}))}^2 \le \frac{C(M, \gamma^+)}{\omega^{1+\gamma^+}} \left(\frac{\omega}{4}\right)^2 \frac{2^{2(n+2)}}{R^2} A_n.$$

Using Theorem 2.11, we get

$$\left(\frac{\omega}{4}\right)^2 \frac{1}{2^{2(n+1)}} A_{n+1} = (k_n - k_{n+1})^2 A_{n+1} \leq \int\int_{Q(R_{n+1}^2, R_{n+1})} (\bar{u}_\omega - k_n)_-^2$$

$$\leq C(d) \left|[\bar{u}_\omega \leq k_n] \cap Q(R_{n+1}^2, R_{n+1})\right|^{\frac{2}{d+2}} \|(\bar{u} - k_n)_-\|_{V^2(Q(R_{n+1}^2, R_{n+1}))}^2$$

$$\leq C(d) A_n^{\frac{2}{d+2}} \|(\bar{u} - k_n)_-\|_{V^2(Q(R_{n+1}^2, R_{n+1}))}^2 ,$$

and consequently

$$A_{n+1} \leq \frac{C(M, d, \gamma^+)}{\omega^{1+\gamma^+}} \frac{2^{4n}}{R^2} A_n^{1+\frac{2}{d+2}}.$$

Defining $Y_n := \frac{A_n}{|Q(R_n^2, R_n)|}$ and noting that

$$\frac{|Q(R_n^2, R_n)|^{1+\frac{2}{d+2}}}{|Q(R_{n+1}^2, R_{n+1})|} \leq 2^{d+4} R^2,$$

we arrive at the algebraic inequality

$$Y_{n+1} \leq \frac{C(M, d, \gamma^+)}{\omega^{1+\gamma^+}} 2^{4n} Y_n^{1+\frac{2}{d+2}}.$$

Now, by Lemma 2.9, if

$$Y_0 \leq C(M, d, \gamma^+)^{-\frac{d+2}{2}} 2^{-(d+2)^2} \omega^{(1+\gamma^+)\frac{d+2}{2}}$$

then $Y_n \to 0$ as $n \to \infty$. Taking

$$\nu_0 := C(M, d, \gamma^+)^{-\frac{d+2}{2}} 2^{-(d+2)^2} \omega^{(1+\gamma^+)\frac{d+2}{2}} \tag{6.10}$$

the above inequality is valid since it is no other than our hypothesis. Since

$$R_n \searrow \frac{R}{2}, \qquad k_n \searrow \frac{\omega}{4},$$

and $Y_n \to 0$ as $n \to \infty$ implies that $A_n \to 0$ as $n \to \infty$, we obtain

$$\left|(x, z) \in Q\left(\left(\frac{R}{2}\right)^2, \frac{R}{2}\right) : \bar{u}_\omega(x, z) \leq \frac{\omega}{4}\right| = 0,$$

that is,

$$u(x, t) > \frac{\omega}{4}, \quad \text{a.e. } (x, t) \in Q\left(a_0 \left(\frac{R}{2}\right)^2, \frac{R}{2}\right).$$

$\square$

The reduction of the oscillation of u follows at once.

Corollary 6.6. *There exist constants $\nu_0 \in (0,1)$, depending on the data and w, and $\sigma_0 \in (0,1)$, such that if (6.6) holds then*

$$\underset{Q\left(a_0\left(\frac{R}{2}\right)^2, \frac{R}{2}\right)}{\text{ess osc}} \quad u \le \sigma_0\, w. \tag{6.11}$$

Proof. From Proposition 6.5 we know that

$$\underset{Q\left(a_0\left(\frac{R}{2}\right)^2, \frac{R}{2}\right)}{\text{ess inf}} \quad u(x,t) \ge \frac{w}{4}.$$

Thereby, since $\mu^+ \ge \underset{Q\left(a_0\left(\frac{R}{2}\right)^2, \frac{R}{2}\right)}{\text{ess sup}} u$,

$$\underset{Q\left(a_0\left(\frac{R}{2}\right)^2, \frac{R}{2}\right)}{\text{ess osc}} \quad u \le \mu^+ - \frac{w}{4} = \left(1 - \frac{1}{4}\right) w = \sigma_0\, w.$$

$\square$

Remark 6.7. We comment here on how to extend the proof to the case of a finite $p > \max\{2, d\}$ in assumption (A1). The novelty is on how to bound the term

$$I^* := \int_{-a_0 R_n^2}^{t} \int_{K_{R_n}} |\nabla \gamma|^2 \chi_{[u_w \le k_n]}$$

now that $|\nabla \gamma|$ is not bounded. We use Hölder's inequality with $q = p/2$ to get

$$I^* \le M^2 \int_{-a_0 R_n^2}^{t} \left(\int_{K_{R_n}} \chi_{[u_w \le k_n]} \right)^{1 - \frac{2}{p}},$$

with $M = \|\gamma\|_{L^\infty(0,T;W^{1,p}(\Omega))}$. An expression of this form is treated as in the case of a degenerate PDE with lower-order terms, for the choices

$$q = \frac{2p(1+\kappa)}{p-2}; \quad r = 2(1+\kappa); \quad \kappa = \frac{2(p-d)}{dp}$$

in the notation of [14, page 24].

6.3 Analysis of the Alternative

Now we assume that (6.6) does not hold; therefore, (6.7) is in force. We will show that, in this case, we can get a result similar to (6.11). Remember that ν_0 was already determined and is given by (6.10).

Lemma 6.8. *Assume that (6.7) holds. There exists a time level*

$$t_0 \in \left[-a_0 R^2, -\frac{\nu_0}{2} a_0 R^2 \right] \tag{6.12}$$

such that

$$\left| x \in K_R : u(x,t_0) > \mu^+ - \frac{\omega}{2} \right| < \left(\frac{1 - \nu_0}{1 - \frac{\nu_0}{2}} \right) |K_R|. \tag{6.13}$$

Proof. If not, then

$$\left| (x,t) \in Q\left(a_0 R^2, R\right) : u(x,t) > \mu^+ - \frac{\omega}{2} \right|$$

$$\geq \int_{-a_0 R^2}^{-\frac{\nu_0}{2} a_0 R^2} \left| x \in K_R : u(x,t) > \mu^+ - \frac{\omega}{2} \right| dt$$

$$\geq \left(\frac{1 - \nu_0}{1 - \frac{\nu_0}{2}} \right) |K_R| \left(1 - \frac{\nu_0}{2} \right) a_0 R^2$$

$$= (1 - \nu_0) \left| Q\left(a_0 R^2, R\right) \right|$$

contradicting (6.7). □

Accordingly, at time level t_0, the portion of the cube K_R where $u(x)$ is close to its supremum is small. In what follows we will show that the same happens for all time levels near the top of the cylinder $Q\left(a_0 R^2, R\right)$.

Lemma 6.9. *There exists $s_1 \in \mathbb{N}$, depending on the data and ω, such that, for all $t \in (t_0, 0)$,*

$$\left| x \in K_R : u(x,t) > \mu^+ - \frac{\omega}{2^{s_1}} \right| < \left(1 - \left(\frac{\nu_0}{2} \right)^2 \right) |K_R|. \tag{6.14}$$

Proof. Consider the cylinder $Q(t_0, R)$, the level $k = \mu^+ - \frac{\omega}{2}$ and put

$$c = \frac{\omega}{2^{n_* + 1}} \quad (n_* \text{ to be chosen}).$$

Define

$$u - k \leq H_k^+ := \operatorname*{ess\,sup}_{Q(t_0, R)} (u - k)_+ \leq \frac{\omega}{2}.$$

If $H_k^+ \leq \frac{\omega}{4}$, the result is trivial for the choice $s_1 = 2$. Assuming $H_k^+ > \frac{\omega}{4}$, the logarithmic function ψ^+ given by

$$\psi^+ = \begin{cases} \ln\left(\frac{H_k^+}{H_k^+ - u + k + c} \right) & \text{if } u > k + c \\ \\ 0 & \text{if } u \leq k + c \end{cases}$$

is well defined in $Q(t_0, R)$ and satisfies the inequalities

$$\psi^+ \leq \ln(2^{n_*}) = n_* \ln 2, \quad \text{since} \quad \frac{H_k^+}{H_k^+ - u + k + c} \leq \frac{H_k^+}{c} \leq \frac{\frac{\omega}{2}}{\frac{\omega}{2^{n_* + 1}}} = 2^{n_*}$$

and, for $u \neq k + c$,

$$0 \leq \left(\psi^+\right)' \leq \frac{1}{H_k^+ - u + k + c} \leq \frac{1}{c} = \frac{2^{n_*+1}}{\omega}$$

and

$$\left(\psi^+\right)'' = \left[\left(\psi^+\right)'\right]^2 \geq 0.$$

In the weak formulation (6.2) [to be rigorous we should consider, as before, formulation (6.3), integrate and take the limit as $h \to 0$; to simplify we proceed formally at this stage] consider the integration over $K_R \times (t_0, t)$, with $t \in (t_0, 0)$, and take $\phi = 2\psi^+ \left(\psi^+\right)' \xi^2$, where $x \to \xi(x)$ is a smooth cutoff function defined in K_R and verifying

$$\begin{cases} 0 \leq \xi \leq 1 \quad \text{in} \quad K_R; \\ \xi \equiv 1 \quad \text{in} \quad K_{(1-\sigma)R}, \quad \text{for some } \sigma \in (0,1); \\ |\nabla \xi| \leq \frac{C}{\sigma R}. \end{cases}$$

Then, for all $t \in (t_0, 0)$,

$$0 = \int_{t_0}^t \int_{K_R} u_t \, 2\psi^+ \left(\psi^+\right)' \xi^2 + \int_{t_0}^t \int_{K_R} u^{\gamma(x,t)} \nabla u \cdot \nabla \left(2\psi^+ \left(\psi^+\right)' \xi^2\right)$$
$$=: J_1 + J_2.$$

The two integrals can be estimated as follows:

$$J_1 = \int_{K_R \times \{t\}} \left(\psi^+\right)^2 \xi^2 - \int_{K_R \times \{t_0\}} \left(\psi^+\right)^2 \xi^2$$
$$\geq \int_{K_R \times \{t\}} \left(\psi^+\right)^2 \xi^2 - n_*^2 \ln^2 2 \left(\frac{1 - \nu_0}{1 - \frac{\nu_0}{2}}\right) |K_R|,$$

using the estimate for ψ^+ and Lemma 6.8;

$$J_2 = \int_{t_0}^t \int_{K_R} u^{\gamma(x,t)} |\nabla u|^2 \, 2 \left(1 + \psi^+\right) \left[\left(\psi^+\right)'\right]^2 \xi^2$$
$$+ 2 \int_{t_0}^t \int_{K_R} u^{\gamma(x,t)} \nabla u \cdot \nabla \xi \, 2\psi^+ \left(\psi^+\right)' \xi$$
$$\geq - \int_{t_0}^t \int_{K_R} u^{\gamma(x,t)} |\nabla \xi|^2 2\psi^+,$$

using Cauchy's inequality. Putting these estimates together, using the bounds for $|\nabla \xi|$ and ψ^+, and recalling the values of t_0 and a_0, we arrive at

$$\int_{K_R \times \{t\}} (\psi^+)^2 \, \xi^2 \leq \left[n_*^2 \ln^2 2 \left(\frac{1 - \nu_0}{1 - \frac{\nu_0}{2}} \right) + 2 n_* \ln 2 \, \frac{C}{\sigma^2 R^2} (-t_0) \right] |K_R|$$

$$\leq \left[n_*^2 \ln^2 2 \left(\frac{1 - \nu_0}{1 - \frac{\nu_0}{2}} \right) + 2 n_* \ln 2 \, \frac{C}{\sigma^2} a_0 \right] |K_R|$$

$$\leq \left[n_*^2 \ln^2 2 \left(\frac{1 - \nu_0}{1 - \frac{\nu_0}{2}} \right) + 2 n_* \ln 2 \, \frac{C(\gamma^+)}{\sigma^2 \omega^{\gamma^+}} \right] |K_R| \, ,$$

valid for all $t \in (t_0, 0)$.

The left hand side is estimated from below integrating over the smaller set

$$S = \left\{ x \in K_{(1-\sigma)R} \, : \, u(x,t) > \mu^+ - \frac{\omega}{2^{n_*+1}} \right\}.$$

On S, $\xi \equiv 1$ and $\psi^+ \geq (n_* - 1) \ln 2$, because

$$\frac{H_k^+}{H_k^+ - u + k + c} \geq \frac{H_k^+}{H_k^+ - \frac{\omega}{2} + \frac{\omega}{2^{n_*}}} = \frac{H_k^+ - \frac{\omega}{2} + \frac{\omega}{2}}{H_k^+ - \frac{\omega}{2} + \frac{\omega}{2^{n_*}}} \geq 2^{n_* - 1},$$

since one has $H_k^+ - \frac{\omega}{2} \leq 0$ and $\frac{\omega}{2} > \frac{\omega}{2^{n_*}}$, $\forall n_* > 1$. Therefore, for all $t \in (t_0, 0)$,

$$|S| \leq \left\{ \left(\frac{n_*}{n_* - 1} \right)^2 \left(\frac{1 - \nu_0}{1 - \frac{\nu_0}{2}} \right) + \frac{C}{\sigma^2 n_* \omega^{\gamma^+}} \right\} |K_R| \, .$$

Consequently, for all $t \in (t_0, 0)$,

$$\left| x \in K_R \, : \, u(x,t) > \mu^+ - \frac{\omega}{2^{n_*+1}} \right| \leq |S| + d\sigma \, |K_R|$$

$$\leq \left\{ \left(\frac{n_*}{n_* - 1} \right)^2 \left(\frac{1 - \nu_0}{1 - \frac{\nu_0}{2}} \right) + \frac{C}{\sigma^2 n_* \omega^{\gamma^+}} + d\sigma \right\} |K_R| \, .$$

The proof is complete once we choose σ so small that $d\sigma \leq \frac{3}{8} \nu_0^2$, then n_* so large that

$$\frac{C}{n_* \sigma^2 \omega^{\gamma^+}} \leq \frac{3}{8} \nu_0^2 \qquad \text{and} \qquad \left(\frac{n_*}{n_* - 1} \right)^2 \leq \left(1 - \frac{\nu_0}{2} \right) (1 + \nu_0) =: \beta > 1,$$

and finally take $s_1 = n_* + 1$.   □

Remark 6.10. Note that, from the choice of σ, we get $\sigma < \frac{3}{8d} \nu_0^2$ and from the two conditions on n_* we obtain

$$n_* \geq \max \left\{ C \nu_0^{-6} \omega^{-\gamma^+} \, ; \, \frac{4}{\nu_0^2} + 2 \right\}.$$

Clearly the number s_1 depends on the data as well as on ω.

Recalling that $t_0 \in \left[-a_0 R^2, -\frac{\nu_0}{2} a_0 R^2\right]$, the next result follows from the previous lemma.

Corollary 6.11. *There exists $s_1 \in \mathbb{N}$, depending on the data and ω, such that, for all $t \in \left(-\frac{\nu_0}{2} a_0 R^2, 0\right)$,*

$$\left| x \in K_R : u(x,t) > \mu^+ - \frac{\omega}{2^{s_1}} \right| < \left(1 - \left(\frac{\nu_0}{2}\right)^2\right) |K_R|. \tag{6.15}$$

The conclusion of Corollary 6.11 will be employed to deduce that, within the cylinder $Q\left(\frac{\nu_0}{2} a_0 R^2, R\right)$, the set where u is close to its supremum is arbitrarily small.

Lemma 6.12. *For all $\nu \in (0,1)$ there exists $s_1 < s_2 \in \mathbb{N}$, depending on the data and ω, such that*

$$\left| (x,t) \in Q\left(\frac{\nu_0}{2} a_0 R^2, R\right) : u(x,t) > \mu^+ - \frac{\omega}{2^{s_2}} \right| \leq \nu \left| Q\left(\frac{\nu_0}{2} a_0 R^2, R\right) \right|.$$

Proof. Consider the cylinder $Q\left(\nu_0 a_0 R^2, 2R\right)$ and the levels $k = \mu^+ - \frac{\omega}{2^s}$, for $s \geq s_1$. In order to obtain energy estimates for the functions $(u-k)_+$ over this cylinder, we take $\phi = (u-k)_+ \xi^2$ in (6.2), where $0 \leq \xi \leq 1$ is a smooth cutoff function defined in $Q\left(\nu_0 a_0 R^2, 2R\right)$ and satisfying

$$\begin{cases} \xi \equiv 0 \quad \text{on} \quad \partial_p Q\left(\nu_0 a_0 R^2, 2R\right); \\ \xi \equiv 1 \quad \text{in} \quad Q\left(\frac{\nu_0}{2} a_0 R^2, R\right); \\ |\nabla \xi| \leq \frac{1}{R}; \qquad 0 \leq \xi_t \leq \frac{1}{\frac{\nu_0}{2} a_0 R^2}. \end{cases}$$

Then, for $t \in \left(-\nu_0 a_0 R^2, 0\right)$, we obtain (again formally)

$$\int_{-\nu_0 a_0 R^2}^{t} \int_{K_{2R}} u_t \, (u-k)_+ \xi^2 + \int_{-\nu_0 a_0 R^2}^{t} \int_{K_{2R}} u^{\gamma(x,t)} \nabla u \cdot \nabla \left((u-k)_+ \xi^2\right) = 0.$$

Now, since $(u-k)_+ \leq \frac{\omega}{2^s}$,

$$\int_{-\nu_0 a_0 R^2}^{t} \int_{K_{2R}} u_t \, (u-k)_+ \xi^2 = \frac{1}{2} \int_{K_{2R} \times \{t\}} (u-k)_+^2 \xi^2$$

$$- \int_{-\nu_0 a_0 R^2}^{t} \int_{K_{2R}} (u-k)_+^2 \xi \xi_t$$

$$\geq - \left(\frac{\omega}{2^s}\right)^2 \frac{1}{\frac{\nu_0}{2} a_0 R^2} \int_{-\nu_0 a_0 R^2}^{t} \int_{K_{2R}} \chi_{[u>k]}$$

and

$$\int_{-\nu_0 a_0 R^2}^t \int_{K_{2R}} u^{\gamma(x,t)} \nabla u \cdot \nabla \left((u-k)_+ \xi^2\right)$$

$$\geq \frac{1}{2a_0} \int_{-\nu_0 a_0 R^2}^t \int_{K_{2R}} |\nabla(u-k)_+|^2 \xi^2 - 2\left(\frac{\omega}{2^s}\right)^2 \frac{1}{R^2} \int_{-\nu_0 a_0 R^2}^t \int_{K_{2R}} \chi_{[u>k]},$$

using Cauchy's inequality with $\epsilon = \frac{1}{4}$ and the fact that when $(u-k)_+$ is not zero, then

$$u > k = \mu^+ - \frac{\omega}{2^s} > \mu^+ - \frac{\omega}{2} = \frac{\omega}{2} > \frac{\omega}{4},$$

since $s \geq s_1 > 1$, and therefore

$$1 \geq u^{\gamma(x,t)} \geq \left(\frac{\omega}{4}\right)^{\gamma^+} = \frac{1}{a_0}.$$

We then have

$$\frac{1}{2a_0} \iint_{Q\left(\frac{\nu_0}{2} a_0 R^2, R\right)} |\nabla(u-k)_+|^2$$

$$\leq \left(\frac{\omega}{2^s}\right)^2 \frac{1}{R^2} \left(\frac{1}{\frac{\nu_0}{2} a_0} + 2\right) \iint_{Q(\nu_0 a_0 R^2, 2R)} \chi_{[u>k]}$$

$$\leq \left(\frac{\omega}{2^s}\right)^2 \frac{1}{R^2} \left(\frac{1}{\frac{\nu_0}{2} a_0} + 2\right) 2^{d+1} \left|Q\left(\frac{\nu_0}{2} a_0 R^2, R\right)\right|$$

and consequently, multiplying by $2a_0$,

$$\iint_{Q\left(\frac{\nu_0}{2} a_0 R^2, R\right)} |\nabla(u-k)_+|^2 \leq \left(\frac{\omega}{2^s}\right)^2 \frac{1}{R^2} \left(\frac{4}{\nu_0} + 4a_0\right) 2^{d+1} \left|Q\left(\frac{\nu_0}{2} a_0 R^2, R\right)\right|$$

$$= \frac{2^{d+3}}{\nu_0} \left(\frac{\omega}{2^s}\right)^2 \frac{1}{R^2} (1 + a_0 \nu_0) \left|Q\left(\frac{\nu_0}{2} a_0 R^2, R\right)\right|$$

$$\leq \frac{C(M, d, \gamma^+)}{\omega^{(1+\gamma^+)\frac{(d+2)}{2}}} \left(\frac{\omega}{2^s}\right)^2 \frac{1}{R^2} \left|Q\left(\frac{\nu_0}{2} a_0 R^2, R\right)\right|,$$

recalling the definition of ν_0 given by (6.10).

Now we consider the levels

$$k_2 = \mu^+ - \frac{\omega}{2^{s+1}} > k_1 = \mu^+ - \frac{\omega}{2^s}, \qquad s = s_1, \ldots, s_2 - 1,$$

and define, for $t \in \left(-\frac{\nu_0}{2} a_0 R^2, 0\right)$,

$$A_s(t) := \left\{ x \in K_R : u(x,t) > \mu^+ - \frac{\omega}{2^s} \right\}$$

and

$$A_s := \int_{-\frac{\nu_0}{2} a_0 R^2}^{0} |A_s(t)| \, dt.$$

Using Lemma 2.7, applied to the function $u(\cdot, t)$ for all times $t \in \left(-\frac{\nu_0}{2} a_0 R^2, 0\right)$, we get

$$\left(\frac{\omega}{2^{s+1}}\right) |A_{s+1}(t)| \leq C(d) \frac{R^{d+1}}{|K_R \setminus A_s(t)|} \int_{[k_1 < u < k_2]} |\nabla u|.$$

Since $\mu^+ - \frac{\omega}{2^s} \geq \mu^+ - \frac{\omega}{2^{s_1}}$, for $s \geq s_1$,

$$|A_s(t)| \leq |A_{s_1}(t)| < \left(1 - \left(\frac{\nu_0}{2}\right)^2\right) |K_R|, \qquad \forall t \in \left(-\frac{\nu_0}{2} a_0 R^2, 0\right),$$

by virtue of (6.15). Then, for every $t \in \left(-\frac{\nu_0}{2} a_0 R^2, 0\right)$,

$$\left(\frac{\omega}{2^{s+1}}\right) |A_{s+1}(t)| \leq \frac{C(d)}{\nu_0^2} R \int_{[k_1 < u < k_2]} |\nabla u|$$

and finally, integrating in time over $\left(-\frac{\nu_0}{2} a_0 R^2, 0\right)$ and using Hölder's inequality, we arrive at

$$\left(\frac{\omega}{2^{s+1}}\right) A_{s+1} \leq \frac{C(d)}{\nu_0^2} R \int_{-\frac{\nu_0}{2} a_0 R^2}^{0} \int_{[k_1 < u < k_2]} |\nabla u|$$

$$\leq \frac{C(d)}{\nu_0^2} R \left(\iint_{Q\left(\frac{\nu_0}{2} a_0 R^2, R\right)} |\nabla (u - k)_+|^2\right)^{\frac{1}{2}} |A_s \setminus A_{s+1}|^{\frac{1}{2}}.$$

According to the previous energy estimates we get, for $s = s_1, \ldots, s_2 - 1$,

$$A_{s+1}^2 \leq \frac{C(M, d, \gamma^+)}{\omega^{\frac{5}{2}(1+\gamma^+)(d+2)}} \left|Q\left(\frac{\nu_0}{2} a_0 R^2, R\right)\right| |A_s \setminus A_{s+1}|,$$

and we then add these inequalities for $s = s_1, \ldots, s_2 - 1$.

Since $\mu^+ - \frac{\omega}{2^{s+1}} \leq \mu^+ - \frac{\omega}{2^{s_2}}$, $A_{s+1} \geq A_{s_2}$, and therefore

$$\sum_{s=s_1}^{s_2 - 1} A_{s+1}^2 \geq (s_2 - s_1) A_{s_2}^2.$$

Note also that $\displaystyle\sum_{s=s_1}^{s_2 - 1} |A_s \setminus A_{s+1}| \leq \left|Q\left(\frac{\nu_0}{2} a_0 R^2, R\right)\right|$. Collecting results, we arrive at

$$A_{s_2}^2 \leq \frac{C(M, d, \gamma^+)}{\omega^{\frac{5}{2}(1+\gamma^+)(d+2)} (s_2 - s_1)} \left|Q\left(\frac{\nu_0}{2} a_0 R^2, R\right)\right|^2$$

and the proof is complete if we choose $s_1 < s_2 \in \mathbb{N}$ sufficiently large so that

$$\frac{C(M, d, \gamma^+)}{\omega^{\frac{5}{2}(1+\gamma^+)(d+2)}(s_2 - s_1)} \leq \nu^2.$$

□

Lemma 6.13. *The number $\nu \in (0, 1)$ can be chosen (and consequently, so can s_2) such that*

$$u(x, t) \leq \mu^+ - \frac{\omega}{2^{s_2+1}}, \quad \text{a.e. } (x, t) \in Q\left(\frac{\nu_0}{2} a_0 \left(\frac{R}{2}\right)^2, \frac{R}{2}\right).$$

Proof. Define two sequences of positive real numbers

$$R_n = \frac{R}{2} + \frac{R}{2^{n+1}}, \qquad k_n = \mu^+ - \frac{\omega}{2^{s_2+1}} - \frac{\omega}{2^{s_2+1+n}}, \quad n = 0, 1, 2, \ldots$$

and construct the family of nested and shrinking cylinders

$$Q_n = Q\left(\frac{\nu_0}{2} a_0 R_n^2, R_n\right).$$

Consider the function $u_\omega = \min\left\{u, \mu^+ - \frac{\omega}{2^{s_2+1}}\right\}$ and, in the weak formulation (6.2), take

$$\phi = (u_\omega - k_n)_+ \xi_n^2,$$

where $0 \leq \xi_n \leq 1$ are smooth cutoff functions defined in Q_n and verifying

$$\begin{cases} \xi_n \equiv 1 \quad \text{in} \quad Q_{n+1}; \quad \xi_n \equiv 0 \quad \text{on} \quad \partial_p Q_n; \\[2mm] |\nabla \xi_n| \leq \frac{2^{n+2}}{R}; \qquad |\Delta \xi_n| \leq \frac{2^{2(n+2)}}{R^2}; \qquad 0 < (\xi_n)_t \leq \frac{2^{2(n+2)}}{\frac{\nu_0}{2} a_0 R^2}. \end{cases}$$

Then, for $t \in \left(-\frac{\nu_0}{2} a_0 R_n^2, 0\right)$, we have formally (again, to be rigorous we would have to argue with the Steklov averages and pass to the limit in h)

$$0 = \int_{-\frac{\nu_0}{2} a_0 R_n^2}^{t} \int_{K_{R_n}} u_t \left((u_\omega - k_n)_+ \xi_n^2\right)$$

$$+ \int_{-\frac{\nu_0}{2} a_0 R_n^2}^{t} \int_{K_{R_n}} u^{\gamma(x,t)} \nabla u \cdot \nabla \left((u_\omega - k_n)_+ \xi_n^2\right) := I_1 + I_2.$$

Arguing as in the proof of Proposition 6.5 (with the obvious changes), we obtain

$$I_1 \geq \frac{1}{2} \int_{K_{R_n} \times \{t\}} (u_\omega - k_n)_+^2 \xi_n^2 - 3 \left(\frac{\omega}{2^{s_2+1}}\right)^2 \frac{2^{2(n+2)}}{\frac{\nu_0}{2} a_0 R^2} \int_{-\frac{\nu_0}{2} a_0 R_n^2}^{t} \int_{K_{R_n}} \chi_{[u_\omega > k_n]},$$

and

$$I_2 \geq \frac{1}{2} \int_{-\frac{\nu_0}{2} a_0 R_n^2}^{t} \int_{K_{R_n}} u_\omega^{\gamma(x,t)} |\nabla(u_\omega - k_n)_+|^2 \xi_n^2$$

$$-2 \int_{-\frac{\nu_0}{2} a_0 R_n^2}^{t} \int_{K_{R_n}} u_\omega^{\gamma(x,t)} (u_\omega - k_n)_+^2 |\nabla \xi_n|^2$$

$$-2 \left(\frac{\omega}{2^{s_2+1}}\right) \int_{-\frac{\nu_0}{2} a_0 R_n^2}^{t} \int_{K_{R_n}} \left(\int_{\mu^+ - \frac{\omega}{2^{s_2+1}}}^{u} s^{\gamma(x,t)} \, ds\right)$$
$$\left(|\Delta \xi_n| + |\nabla \xi_n|^2\right) \chi_{\left[u \geq \mu^+ - \frac{\omega}{2^{s_2+1}}\right]}$$

$$-\left(\frac{\omega}{2^{s_2+1}}\right) \int_{-\frac{\nu_0}{2} a_0 R_n^2}^{t} \int_{K_{R_n}} \left(\int_{\mu^+ - \frac{\omega}{2^{s_2+1}}}^{u} s^{\gamma(x,t)} (-\ln s) \, ds\right)$$
$$\left(|\nabla \gamma|^2 + |\nabla \xi_n|^2\right) \chi_{\left[u \geq \mu^+ - \frac{\omega}{2^{s_2+1}}\right]},$$

and, ultimately, the estimate

$$\sup_{-\frac{\nu_0}{2} a_0 R_n^2 < t < 0} \int_{K_{R_n} \times \{t\}} (u_\omega - k_n)_+^2 \, \xi_n^2 + \frac{1}{a_0} \iint_{Q_n} |\nabla(u_\omega - k_n)_+|^2 \xi_n^2$$

$$\leq \left(\frac{\omega}{2^{s_2+1}}\right)^2 \frac{2^{2(n+2)}}{R^2} \left\{\frac{6}{\frac{\nu_0}{2} a_0} + \frac{C(M)}{\omega}\right\} \iint_{Q_n} \chi_{[u_\omega \geq k_n]}.$$

Introducing the change of variables

$$z = \frac{t}{\frac{\nu_0}{2} a_0}$$

and defining the new functions

$$\bar{u}_\omega(x, z) = u_\omega\left(x, \frac{\nu_0}{2} a_0 z\right), \qquad \bar{\xi}_n(x, z) = \xi_n\left(x, \frac{\nu_0}{2} a_0 z\right),$$

the previous estimate reads

$$\sup_{-R_n^2 < z < 0} \int_{K_{R_n} \times \{z\}} (\bar{u}_\omega - k_n)_+^2 \, \bar{\xi}_n^2 + \frac{\nu_0}{2} \iint_{Q(R_n^2, R_n)} |\nabla(\bar{u}_\omega - k_n)_+|^2 \bar{\xi}_n^2$$

$$\leq \left(\frac{\omega}{2^{s_2+1}}\right)^2 \frac{2^{2(n+2)}}{R^2} \left\{6 + \frac{C(M)}{\omega} \nu_0 a_0\right\} \iint_{Q(R_n^2, R_n)} \chi_{[\bar{u}_\omega \geq k_n]}.$$

Multiplying the above estimate by $\frac{2}{\nu_0} \geq 1$, we arrive at

$$\|(\bar{u}_\omega - k_n)_+\|_{V^2(Q(R_{n+1}^2, R_{n+1}))}^2 \leq C(M) \left(\frac{\omega}{2^{s_2+1}}\right)^2 \frac{2^{2(n+2)}}{R^2} \left\{\frac{1}{\nu_0} + \frac{a_0}{\omega}\right\} A_n$$

$$\leq \frac{C(M, d, \gamma^+)}{\omega^{(1+\gamma^+)(\frac{d+2}{2})}} \left(\frac{\omega}{2^{s_2+1}}\right)^2 \frac{2^{2(n+2)}}{R^2} A_n,$$

where A_n is defined as

$$A_n := \int_{-R_n^2}^{0} \int_{K_{R_n}} \chi_{[\bar{u}_\omega \geq k_n]}.$$

Now, we first obtain

$$A_{n+1} \leq \frac{C(M, d, \gamma^+)}{\omega^{(1+\gamma^+)(\frac{d+2}{2})}} \frac{2^{4n}}{R^2} A_n^{1+\frac{2}{d+2}},$$

and then, defining $Y_n := \frac{A_n}{|Q(R_n^2, R_n)|}$, we get the algebraic inequality

$$Y_{n+1} \leq \frac{C(M, d, \gamma^+)}{\omega^{(1+\gamma^+)(\frac{d+2}{2})}} 2^{4n} Y_n^{1+\frac{2}{d+2}}.$$

As before, the result is proved if we can assure that

$$Y_0 \leq C(M, d, \gamma^+)^{-\frac{d+2}{2}} \omega^{(1+\gamma^+)\frac{(d+2)^2}{4}} 2^{-(d+2)^2}$$
$$= C(M, d, \gamma^+) \omega^{(1+\gamma^+)\frac{(d+2)^2}{4}} =: \nu.$$

By Lemma 6.12, for this value of ν there exists $s_1 < s_2 \in \mathbb{N}$ such that

$$\frac{\left|(x, z) \in Q\left(R^2, R\right) : \bar{u}(x, z) > \mu^+ - \frac{\omega}{2^{s_2}}\right|}{|Q\left(R^2, R\right)|} \leq \nu$$

which implies $Y_0 \leq \nu$. Then we can conclude that $Y_n \to 0$ when $n \to \infty$, and the result follows. □

Proposition 6.14. *There exist positive numbers $\nu_0, \sigma_1 \in (0, 1)$, depending on the data and on ω, such that, if (6.7) holds true then*

$$\underset{Q\left(\frac{\nu_0}{2} a_0\left(\frac{R}{2}\right)^2, \frac{R}{2}\right)}{\text{ess osc}} u \leq \sigma_1 \omega. \tag{6.16}$$

Proof. The proof is trivial and similar to the proof of Corollary 6.6. We have $\sigma_1 = 1 - \frac{1}{2^{s_2+1}}$.

□

Proof of Proposition 6.3. Recalling the conclusions of Corollary 6.6 and Proposition 6.14, we take

$$\sigma = \max\{\sigma_0, \sigma_1\} = \sigma_1,$$

since $\sigma_0 = 1 - \frac{1}{4} < 1 - \frac{1}{2^{s_2}+1} = \sigma_1$, because $s_2 > 1$. As $\nu_0 \in (0,1)$,

$$Q\left(\frac{\nu_0}{2}a_0\left(\frac{R}{2}\right)^2, \frac{R}{2}\right) \subset Q\left(a_0\left(\frac{R}{2}\right)^2, \frac{R}{2}\right)$$

and the result follows. □

Remark 6.15. The extension to the singular case

$$-1 < \gamma^- \le \gamma(x,t) \le \gamma^+ < 0$$

is treated in [26].

Phase Transitions: The Doubly Singular Stefan Problem

Another interesting equation is the doubly singular PDE

$$\gamma(u)_t - \operatorname{div} |\nabla u|^{p-2} \nabla u = 0, \qquad 1 < p < 2, \tag{7.1}$$

where γ is a coercive, maximal monotone graph in $\mathbb{R} \times \mathbb{R}$ given by

$$\gamma(s) = \begin{cases} s & \text{if } s < 0 \\ [0, \lambda] & \text{if } s = 0 \\ s + \lambda & \text{if } s > 0, \end{cases} \tag{7.2}$$

with $\lambda > 0$. The equation is to be interpreted in the sense of the graphs, i.e., for a choice $v \in \gamma(u)$, and exhibits a double singularity: as $1 < p < 2$, its modulus of ellipticity $|\nabla u|^{p-2}$ blows up at points where $|\nabla u| = 0$; an additional singularity occurs in the time derivative since, loosely speaking, $\gamma'(0) = \infty$.

Graphs $\gamma(\cdot)$ such as this one, with a single jump at the origin, arise from the weak formulation of the classical Stefan problem (cf. [40]), that corresponds to the case $p = 2$ and models a solid–liquid phase transition (such as water–ice) at constant temperature for a substance obeying Fourier's law. A natural question to ask is whether the transition of phase occurs with a continuous temperature across the water–ice interface. This issue, raised initially by Oleinik in the 1950's, and reported in the book [37], is at the origin of the modern theory of local regularity for solutions of degenerate and/or singular evolution equations. The coercivity of $\gamma(\cdot)$ is essential for a solution to be continuous, as illustrated by examples and counterexamples in [16].

Consider the more general case

$$\gamma(u)_t - \operatorname{div} \mathbf{a}(x, t, u, \nabla u) = b(x, t, u, \nabla u) \qquad \text{in } \mathcal{D}'(\Omega_T), \tag{7.3}$$

with $\mathbf{a}$ and b satisfying the usual structure assumptions for $p = 2$. Solutions of (7.3) are continuous with a modulus of continuity that is not Hölder; this was established in [7] (for the Laplacian), [11], [48], [49], [57], for $\gamma(\cdot)$ exhibiting a single jump. This raises naturally the question of a graph $\gamma(\cdot)$ exhibiting multiple jumps and/or singularities of other nature. For these rather general graphs, in the mid 1990's, it was established in [22] that solutions of (7.3) are continuous provided $d = 2$. For dimension $d \geq 3$ the same conclusion holds provided the principal part of the differential equation is *exactly* the Laplacian. Several recent investigations have extended and improved these results for specific graphs ([23], [24]). However, for $d \geq 3$, it is still an open question whether solutions of (7.3), with its full quasilinear structure (even with $p = 2$) and for a general coercive maximal monotone graph $\gamma(\cdot)$, are continuous in their domain of definition.

7.1 Regularization of the Maximal Monotone Graph

Here, we restrict our attention to (7.1). The analysis we will perform involves regularizing the maximal monotone graph γ and obtaining *a priori* estimates for the regularized problem that are independent of the regularization. The ultimate goal is to use Ascoli's theorem to obtain the continuity of the solution of the original problem by proving it is the limit of a sequence of equibounded and equicontinuous approximate solutions. We will not be concerned with problems of existence of the weak solution for boundary value problems associated with (7.1) or the convergence of the sequence of approximate solutions to the weak solution; this problem was treated in [50] (see also the classical reference [39]).

The consequence of estimating uniformly the regularization of the maximal monotone graph is the appearance of a third power (power 1) in the energy estimates. We will thus be dealing with three powers (1, p and 2) related by $1 < p < 2$. In this case, the price to be paid for the recovering of the homogeneity in the energy estimates is a dependence on the oscillation in the various constants that are determined along the proof. Owing to this fact we will no longer be able to exhibit a modulus of continuity but only to define it implicitly independently of the regularization. This is enough to obtain a continuous solution for the original problem, via Ascoli's theorem, but the Hölder continuity, that holds in the case $\gamma(s) = s$, is lost.

To be precise, consider the maximal monotone graph H associated with the Heaviside function

$$H(s) = \begin{cases} 0 & \text{if } s < 0 \\ [0,1] & \text{if } s = 0 \\ 1 & \text{if } s > 0 \end{cases}$$

and let $\gamma(s) = s + \lambda H(s)$, where λ is a positive constant (physically, the latent heat of the phase transition):

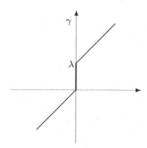

Let $0 < \epsilon \ll 1$ and consider the function

$$\gamma_\epsilon(s) = s + \lambda H_\epsilon(s),$$

where H_ϵ is a C^∞ approximation of the Heaviside function, such that

$$H_\epsilon(s) = 0 \quad \text{if} \quad s \leq 0 \ ; \qquad H_\epsilon(s) = 1 \quad \text{if} \quad s \geq \epsilon,$$

$H'_\epsilon(s) \geq 0$, $s \in \mathbb{R}$ and $H_\epsilon \longrightarrow H$ uniformly in compact subsets of $\mathbb{R} \setminus \{0\}$ as $\epsilon \to 0$. The function γ_ϵ is Lipschitz continuous, together with its inverse, and satisfies

$$1 \leq \gamma'_\epsilon(s) \leq 1 + \lambda L_\epsilon, \quad s \in \mathbb{R}, \qquad (7.4)$$

where $L_\epsilon \equiv \mathcal{O}(\frac{1}{\epsilon})$ is the Lipschitz constant of H_ϵ. Its inverse $\beta_\epsilon = \gamma_\epsilon^{-1}$ satisfies

$$0 < \frac{1}{1 + \lambda L_\epsilon} \leq \beta'_\epsilon(s) \leq 1, \quad s \in \mathbb{R}. \qquad (7.5)$$

Definition 7.1. *An approximate solution of equation* (7.1) *is a function*

$$\theta_\epsilon \in L^\infty(\Omega_T) \cap H^1\left(0, T; L^2(\Omega)\right) \cap L^\infty\left(0, T; W_0^{1,p}(\Omega)\right)$$

such that, for a.e. $t \in (0, T)$,

$$\int_{\Omega \times \{t\}} \left\{ [\gamma_\epsilon(\theta_\epsilon)]_t \ \varphi + |\nabla\theta_\epsilon|^{p-2}\nabla\theta_\epsilon \cdot \nabla\varphi \right\} dx = 0, \qquad (7.6)$$

for all $\varphi \in W_0^{1,p}(\Omega)$.

We assume that the approximate solutions satisfy the uniform bound

$$\|\theta_\epsilon\|_{L^\infty(\Omega_T)} \leq M.$$

It is easily seen that (7.6) corresponds, in the sense of distributions, to the equation

$$u_t - \operatorname{div} |\nabla\beta_\epsilon(u)|^{p-2}\nabla\beta_\epsilon(u) = 0,$$

with $u = \gamma_\epsilon(\theta_\epsilon)$, which is of the form (4.19) and satisfies the structure conditions

$$|\nabla\beta_\epsilon(u)|^{p-2}\nabla\beta_\epsilon(u) \cdot \nabla u \geq \left(\frac{1}{1+\lambda L_\epsilon}\right)^{p-1}|\nabla u|^p$$

and

$$||\nabla\beta_\epsilon(u)|^{p-2}\nabla\beta_\epsilon(u)| \leq |\nabla u|^{p-1}.$$

As a consequence of this fact, using the general theory, the approximate solutions are Hölder continuous. But the constant

$$C_0(\epsilon) := \left(\frac{1}{1+\lambda L_\epsilon}\right)^{p-1}$$

deteriorates as $\epsilon \to 0$ and whatever is proved using this approach is lost in the limit. What we need to do is to obtain energy estimates that are independent of the approximating parameter ϵ. This will allow us to show that the approximate solutions are continuous independently of the approximation.

7.2 A Third Power in the Energy Estimates

Consider a cylinder $Q(\tau, \rho) \subset \Omega_T$, and a piecewise smooth cutoff function ζ in $Q(\tau, \rho)$ such that

$$0 \leq \zeta \leq 1; \qquad |\nabla\zeta| < \infty; \qquad \zeta = 0, \; x \notin K_\rho.$$

Let $k < M$ and $\varphi = -(\theta_\epsilon - k)_- \zeta^p$ in (7.6). Integrating in time over $(-\tau, t)$ for $t \in (-\tau, 0)$, the first term gives

$$-\int_{-\tau}^t \int_{K_\rho} [\gamma_\epsilon(\theta_\epsilon)]_t \, (\theta_\epsilon - k)_- \zeta^p$$

$$= \int_{K_\rho} \int_{-\tau}^t \partial_t \left(\int_0^{(\theta_\epsilon-k)_-} \gamma_\epsilon'(k-s)s \, ds\right) \zeta^p$$

$$= \int_{K_\rho \times \{t\}} \left(\int_0^{(\theta_\epsilon-k)_-} \gamma_\epsilon'(k-s)s \, ds\right) \zeta^p$$

$$- \int_{K_\rho \times \{-\tau\}} \left(\int_0^{(\theta_\epsilon-k)_-} \gamma_\epsilon'(k-s)s \, ds\right) \zeta^p$$

$$- p \int_{-\tau}^t \int_{K_\rho} \left(\int_0^{(\theta_\epsilon-k)_-} \gamma_\epsilon'(k-s)s \, ds\right) \zeta^{p-1}\zeta_t$$

$$\geq \frac{1}{2} \int_{K_\rho \times \{t\}} (\theta_\epsilon - k)_-^2 \zeta^p - 2(M+\lambda) \int_{K_\rho \times \{-\tau\}} (\theta_\epsilon - k)_- \zeta^p$$

$$- 2p(M+\lambda) \int_{-\tau}^t \int_{K_\rho} (\theta_\epsilon - k)_- \zeta^{p-1}\zeta_t, \tag{7.7}$$

since we have, recalling (7.4),

$$\int_0^{(\theta_\epsilon - k)_-} \gamma_\epsilon'(k-s)s\, ds \geq \int_0^{(\theta_\epsilon - k)_-} s\, ds = \frac{1}{2}(\theta_\epsilon - k)_-^2$$

and

$$\int_0^{(\theta_\epsilon - k)_-} \gamma_\epsilon'(k-s)s\, ds \leq (\theta_\epsilon - k)_- \int_0^{(\theta_\epsilon - k)_-} \gamma_\epsilon'(k-s)\, ds$$

$$= (\theta_\epsilon - k)_- \left[\gamma_\epsilon(k) - \gamma_\epsilon(\theta_\epsilon)\right] \leq 2(M + \lambda)\,(\theta_\epsilon - k)_-.$$

Concerning the other term, we have

$$\int_{-\tau}^t \int_{K_\rho} |\nabla\theta_\epsilon|^{p-2}\nabla\theta_\epsilon \cdot \nabla\left[-(\theta_\epsilon - k)_-\zeta^p\right] = \int_{-\tau}^t \int_{K_\rho} |\nabla(\theta_\epsilon - k)_-\zeta|^p$$

$$-p\int_{-\tau}^t \int_{K_\rho} |\nabla\theta_\epsilon|^{p-2}\nabla\theta_\epsilon \cdot \nabla\zeta\left[\zeta^{p-1}(\theta_\epsilon - k)_-\right]$$

$$\geq \frac{1}{2}\int_{-\tau}^t \int_{K_\rho} |\nabla(\theta_\epsilon - k)_-\zeta|^p - C(p)\int_{-\tau}^t \int_{K_\rho} (\theta_\epsilon - k)_-^p |\nabla\zeta|^p, \qquad (7.8)$$

using Young's inequality. Since $t \in (-\tau, 0)$ is arbitrary, we can combine estimates (7.7) and (7.8) to obtain the next result.

Proposition 7.2. *Let θ_ϵ be an approximate solution of (7.1) and let $k < M$. There exists a constant C, that is independent of ϵ, such that for every cylinder $Q(\tau, \rho) \subset \Omega_T$,*

$$\sup_{-\tau < t < 0} \int_{K_\rho \times \{t\}} (\theta_\epsilon - k)_-^2 \zeta^p\, dx + \int_{-\tau}^0 \int_{K_\rho} |\nabla(\theta_\epsilon - k)_-\zeta|^p\, dx\, dt$$

$$\leq C\int_{K_\rho \times \{-\tau\}} (\theta_\epsilon - k)_-\zeta^p\, dx + C\int_{-\tau}^0 \int_{K_\rho} (\theta_\epsilon - k)_-^p |\nabla\zeta|^p\, dx\, dt$$

$$+C\int_{-\tau}^0 \int_{K_\rho} (\theta_\epsilon - k)_-\zeta^{p-1}\zeta_t\, dx\, dt. \qquad (7.9)$$

The conclusion is that dealing with the singularity produced by the maximal monotone graph involves a regularization procedure and the search for estimates that are uniform with respect to that regularization, *i.e.*, that are independent of the parameter ϵ. Since γ_ϵ' is not uniformly bounded above near the singularity, this leads to the appearance of integral terms in the energy estimates involving the L^1-norm of the solution, besides the L^2-norm. Thus, with three powers in play, the inhomogeneity in the estimates becomes more severe. The ultimate effect will be the loss of the Hölder continuity in the limit, for which only continuity will be obtained via Ascoli's theorem.

When $k > \epsilon$, we are above the singularity and the energy estimates for $(\theta_\epsilon - k)_+$ read as follows.

Proposition 7.3. *Let θ_ϵ be an approximate solution of (7.1) and $k > \epsilon$. Then there exists a constant C, that is independent of ϵ, such that for every cylinder $Q(\tau, \rho) \subset \Omega_T$,*

$$\sup_{-\tau < t < 0} \int_{K_\rho \times \{t\}} (\theta_\epsilon - k)_+^2 \zeta^p \, dx + \int_{-\tau}^0 \int_{K_\rho} |\nabla(\theta_\epsilon - k)_+ \zeta|^p \, dx \, dt$$

$$\leq C \int_{K_\rho \times \{-\tau\}} (\theta_\epsilon - k)_+^2 \zeta^p \, dx + C \int_{-\tau}^0 \int_{K_\rho} (\theta_\epsilon - k)_+^p |\nabla \zeta|^p \, dx \, dt$$

$$+ C \int_{-\tau}^0 \int_{K_\rho} (\theta_\epsilon - k)_+^2 \zeta^{p-1} \zeta_t \, dx \, dt. \tag{7.10}$$

Concerning the logarithmic estimates, we obtain the following results. The reasoning is that of Section 2.3.

Proposition 7.4. *Let θ_ϵ be an approximate solution of (7.1), $k \in \mathbb{R}$ and $0 < c < H_{\theta_\epsilon, k}^-$. There exists a constant $C > 0$, that is independent of ϵ, such that for every cylinder $Q(\tau, \rho) \subset \Omega_T$,*

$$\sup_{-\tau < t < 0} \int_{K_\rho \times \{t\}} \left[\psi^-(\theta_\epsilon) \right]^2 \zeta^p \, dx$$

$$\leq \int_{K_\rho \times \{-\tau\}} \left(\int_k^{\theta_\epsilon} 2\gamma_\epsilon'(s) \psi^-(s) (\psi^-)'(s) \, ds \right)_+ \zeta^p \, dx$$

$$+ C \int_{-\tau}^0 \int_{K_\rho} \psi^-(\theta_\epsilon) |(\psi^-)'(\theta_\epsilon)|^{2-p} |\nabla \zeta|^p \, dx \, dt. \tag{7.11}$$

Remark 7.5. In this estimate there is a dependence on ϵ through γ_ϵ'. We will see later how to overcome this difficulty.

Proposition 7.6. *Let θ_ϵ be an approximate solution of (7.1), $k > \epsilon$ and $0 < c < H_{\theta_\epsilon, k}^+$. There exists a constant $C > 0$, that is independent of ϵ, such that for every cylinder $Q(\tau, \rho) \subset \Omega_T$,*

$$\sup_{-\tau < t < 0} \int_{K_\rho \times \{t\}} \left[\psi^+(\theta_\epsilon) \right]^2 \zeta^p \, dx \leq \int_{K_\rho \times \{-\tau\}} \left[\psi^+(\theta_\epsilon) \right]^2 \zeta^p \, dx$$

$$+ C \int_{-\tau}^0 \int_{K_\rho} \psi^+(\theta_\epsilon) \left[(\psi^+)'(\theta_\epsilon) \right]^{2-p} |\nabla \zeta|^p \, dx \, dt. \tag{7.12}$$

7.3 The Intrinsic Geometry

The study of the interior regularity of the approximate solution, namely showing that it is continuous independently of ϵ, requires the choice of the right intrinsic geometry that somehow reflects the two singularities in the equation.

To fully understand what is at stake, let us observe that a bridge between the singularity in time and a degeneracy in space can be made through rewriting equation (7.1) in terms of a $v \in \gamma(u)$, taking into account that $u = \gamma^{-1}(v)$ and γ^{-1} is now a well defined function, such that $\gamma^{-1}(s) = 0$ for $0 \le s \le \lambda$. It is clear that the time singularity in the u-equation becomes a space degeneracy for the v-equation. In the case of equation (7.1) with $p > 2$, which was treated in [51], we are in the presence of two types of degeneracy in the principal part of the equation and that explains why a rescaling in time is enough. Also, in the case $1 < p < 2$ but with no jumps (i.e., for $\gamma(s) = s$), there is only a singularity, so a rescaling in space suffices (see [14, Ch. 4]). Here, we have the equivalent of both a singularity and a degeneracy in the principal part and so we need both rescalings, in space and in time.

In order to simplify the notation, from now on we drop the subscript ϵ in θ_ϵ. Given $R > 0$, sufficiently small such that

$$Q\left(R, R^{\frac{1}{2}}\right) \subset \Omega_T,$$

define

$$\mu^- := \operatorname*{ess\,inf}_{Q(R,R^{\frac{1}{2}})} \theta \; ; \quad \mu^+ := \operatorname*{ess\,sup}_{Q(R,R^{\frac{1}{2}})} \theta \; ; \quad \omega := \operatorname*{ess\,osc}_{Q(R,R^{\frac{1}{2}})} \theta = \mu^+ - \mu^-$$

and construct the cylinder

$$Q\left(a_0 R^p, c_0 R\right), \qquad a_0 = \left(\frac{\omega}{A}\right)^{(1-p)(2-p)} \; ; \qquad c_0 = \left(\frac{\omega}{B}\right)^{p-2}, \qquad (7.13)$$

where $A = 2^s$ and $B = 2^{\bar{s}}$, for some $s, \bar{s} > 1$ to be chosen. Observe that for $p = 2$, $a_0 = c_0 = 1$, and these are the standard parabolic cylinders, reflecting the natural homogeneity of the space and time variables.

We assume, without loss of generality, that $\omega \le 1$ and also that

$$\omega > \max\left\{ AR^{\frac{1}{2-p}}, \; BR^{\frac{1}{2(2-p)}} \right\}. \qquad (7.14)$$

Then $Q(a_0 R^p, c_0 R) \subset Q(R, R^{\frac{1}{2}})$ and

$$\operatorname*{ess\,osc}_{Q(a_0 R^p, c_0 R)} \theta \le \omega. \qquad (7.15)$$

We now consider the cube $K_{c_0 R}$ partitioned in disjoint subcubes, each congruent to $K_{d_* R}$, with $d_* = \left(\frac{\omega}{2^{n_*+1}}\right)^{p-2}$, for n_* to be determined:

$$[\bar{x} + K_{d_* R}], \qquad \bar{x} \in K_{\mathcal{R}(\omega)}, \qquad \mathcal{R}(\omega) := c_0 R - d_* R = L_1 d_* R$$

$$L_1 = \left(\frac{B}{2^{n_*+1}}\right)^{2-p} - 1, \qquad B > 2^{n_*+1}.$$

Since we may arrange L_1 to be an integer, we can look at $K_{c_0 R}$ as the disjoint union, up to a set of measure zero, of L_1^d cubes of the above type. Then we may regard the cylinder $Q(a_0 R^p, c_0 R)$ as the disjoint union, up to a set of measure zero, of L_1^d subcylinders of the type $[(\bar{x}, 0) + Q(a_0 R^p, d_* R)]$.

We next consider subcylinders of $[(\bar{x}, 0) + Q(a_0 R^p, d_* R)]$ of the form

$$[(\bar{x}, \bar{t}) + Q(d^* R^p, d_* R)], \qquad d^* = \left(\frac{\omega}{2}\right)^{(1-p)(2-p)},$$

which are contained in $[(\bar{x}, 0) + Q(a_0 R^p, d_* R)]$ assuming that $A > 2$ and

$$\bar{t} \in \mathcal{I}(\omega) := (-a_0 R^p + d^* R^p, 0). \tag{7.16}$$

The proof of the equicontinuity relies on the study of two complementary cases and the achievement of the same conclusion for both, namely the reduction of the oscillation. For a given constant $\nu_0 \in (0, 1)$, which will be determined later only in terms of the data and ω, we assume that either

The First Alternative

there exists $\bar{t} \in \mathcal{I}(\omega)$ such that, for all $\bar{x} \in K_{\mathcal{R}(\omega)}$,

$$\frac{\left|(x, t) \in [(\bar{x}, \bar{t}) + Q(d^* R^p, d_* R)] \; : \; \theta(x, t) < \mu^- + \frac{\omega}{2}\right|}{|Q(d^* R^p, d_* R)|} \leq \nu_0 \tag{7.17}$$

or

The Second Alternative

for every $\bar{t} \in \mathcal{I}(\omega)$, there exists $\bar{x} \in K_{\mathcal{R}(\omega)}$ such that

$$\frac{\left|(x, t) \in [(\bar{x}, \bar{t}) + Q(d^* R^p, d_* R)] \; : \; \theta(x, t) > \mu^+ - \frac{\omega}{2}\right|}{|Q(d^* R^p, d_* R)|} < 1 - \nu_0. \tag{7.18}$$

In the first part of the alternative, we deal with the singularity in time (degeneracy in space) so what dominates the geometry is the scaling in time; the type of partition of the cylinders that is considered is a reflection of this fact. In the second part of the alternative, everything takes place above the singularity in time, the singular character of the principal part thus being the dominant factor; it comes with no surprise that the type of partition considered there is a partition in space.

As a consequence of the alternative, we obtain a constant $\sigma = \sigma(\omega) \in (0, 1)$, depending only on the data and ω, such that

$$\operatorname*{ess\,osc}_{Q(d^*(\frac{R}{8})^p, c_0 \frac{R}{8})} \theta \leq \sigma(\omega)\omega.$$

As usually, we next define recursively two sequences of real positive numbers. Let

$$\omega_1 = \sigma(\omega)\omega \qquad \text{and} \qquad R_1 = \frac{R}{C(\omega_1)},$$

where

$$C(\omega_1) = \left(\frac{A}{2}\right)^{\frac{(p-1)(2-p)}{p}} 8\sigma(\omega)^{\frac{(1-p)(2-p)}{p}} B(\omega_1)^{2-p}\sigma(\omega)^{\frac{p-2}{p}} > 8.$$

Defining

$$Q_1 = (a_1 R_1^p, c_1 R_1) ; \qquad \text{with} \quad a_1 = \left(\frac{\omega_1}{A}\right)^{(1-p)(2-p)} , \qquad c_1 = \left(\frac{\omega_1}{B(\omega_1)}\right)^{p-2}$$

and noting that

$$\begin{aligned}
a_1 R_1^p &= \left(\frac{\omega_1}{A}\right)^{(1-p)(2-p)} \frac{R^p}{C(\omega_1)^p} \\
&= \left(\frac{\omega}{2}\right)^{(1-p)(2-p)} \left(\frac{R}{8}\right)^p \frac{\sigma(\omega)^{2-p}}{B(\omega_1)^{p(2-p)}} \\
&\leq d^* \left(\frac{R}{8}\right)^p
\end{aligned}$$

and

$$\begin{aligned}
c_1 R_1 &= \left(\frac{\omega_1}{B(\omega_1)}\right)^{p-2} \frac{R}{C(\omega_1)} \\
&= \left(\frac{\omega}{B}\right)^{p-2} \left(\frac{R}{8}\right) \frac{1}{B^{2-p}\left(\frac{A}{2}\right)^{\frac{(p-1)(2-p)}{p}}} \\
&\leq c_0 \frac{R}{8}
\end{aligned}$$

we get

$$Q_1 \subset Q\left(d^*\left(\frac{R}{8}\right)^p, c_0\frac{R}{8}\right)$$

and, consequently,

$$\underset{Q_1}{\text{ess osc}} \; \theta \leq \underset{Q\left(d^*\left(\frac{R}{8}\right)^p, c_0\frac{R}{8}\right)}{\text{ess osc}} \; \theta \leq \sigma(\omega)\omega = \omega_1.$$

The process can now be repeated starting from Q_1 since (7.15) holds in this cylinder. We then define the following recursive sequences of real positive numbers

$$\begin{cases} \omega_0 = \omega \\ \omega_{n+1} = \sigma(\omega_n)\omega_n \end{cases} \qquad \text{and} \qquad \begin{cases} R_0 = R \\ R_{n+1} = \frac{R_n}{C(\omega_{n+1})} \end{cases}$$

and construct the family of nested and shrinking cylinders

$$Q_n = (a_n R_n^p, c_n R_n), \qquad n = 0, 1, \ldots$$

with

$$a_n = \left(\frac{\omega_n}{A}\right)^{(1-p)(2-p)}, \quad c_n = \left(\frac{\omega_n}{B(\omega_n)}\right)^{p-2}.$$

Theorem 7.7. *The sequences $(\omega_n)_n$ and $(R_n)_n$ are decreasing sequences converging to zero. Moreover, for every $n = 0, 1, \ldots$*

$$Q_{n+1} \subset Q_n \qquad \text{and} \qquad \underset{Q_n}{\text{ess osc}}\ \theta \le \omega_n. \qquad (7.19)$$

Proof. The sequences are obviously decreasing and bounded below by zero, so to show that they converge to zero we just need to show that they cannot converge to a positive number. As far as $(R_n)_n$ is concerned this conclusion follows immediately from

$$\frac{R_{n+1}}{R_n} = \frac{1}{C(\omega_{n+1})} < \frac{1}{8}.$$

With $(\omega_n)_n$ the situation is more delicate since

$$\frac{\omega_{n+1}}{\omega_n} = \sigma(\omega_n) \nearrow 1,$$

due to the special form of σ (see Corollaries 7.13 and 7.23). So we suppose that $\omega_n \searrow \alpha > 0$ and observe that, in that case,

$$\sigma(\omega_n) \nearrow \sigma(\alpha) < 1.$$

Consequently, $\omega_{n+1} = \sigma(\omega_n)\, \omega_n \le \sigma(\alpha)\, \omega_n$ and

$$\omega_n \le [\sigma(\alpha)]^n\, \omega, \qquad n = 0, 1, \ldots,$$

which implies that $\omega_n \to 0$, a contradiction.

Relations (7.19) follow at once from the recursive process used to define the sequences. □

The next result is now standard (see Section 4.4).

Theorem 7.8. *The sequence $(\theta_\epsilon)_\epsilon$ is locally equicontinuous, i.e., for each θ_ϵ, there exists an interior modulus of continuity that is independent of ϵ. Therefore equation (7.1) has, at least, one locally continuous solution.*

7.4 Analyzing the Singularity in Time

Assume that (7.17) holds in $[(\bar{x}, \bar{t}) + Q(d^* R^p, d_* R)]$, where $[\bar{x} + K_{d_* R}]$ is any subcube of the partition of $K_{c_0 R}$ and $\bar{t}$ is the same in all these cylinders. Since $K_{c_0 R}$ is the disjoint union, up to a set of measure zero, of subcubes of the form $[\bar{x} + K_{d_* R}]$, the first alternative implies that there exists a cylinder of the type $[(0, \bar{t}) + Q(d^* R^p, c_0 R)]$ in which

$$\frac{\left|(x, t) \in [(0, \bar{t}) + Q(d^* R^p, c_0 R)] \;:\; \theta(x, t) < \mu^- + \frac{\omega}{2}\right|}{|Q(d^* R^p, c_0 R)|} \leq \nu_0. \tag{7.20}$$

Lemma 7.9. *There exists a constant $\nu_0 \in (0, 1)$, depending only on the data and ω, such that if (7.17) holds for some $\bar{t} \in \mathcal{I}(\omega)$ and all $\bar{x} \in K_{\mathcal{R}(\omega)}$,*

$$\theta(x, t) > \mu^- + \frac{\omega}{4}, \quad a.e. \ (x, t) \in \left[(0, \bar{t}) + Q\left(d^* \left(\frac{R}{2}\right)^p, c_0 \frac{R}{2}\right)\right]. \tag{7.21}$$

Proof. According to the remark preceding the statement of the lemma, (7.20) is in force. After translation, we may assume that $\bar{t} = 0$. Define two decreasing sequences of numbers

$$R_n = \frac{R}{2} + \frac{R}{2^{n+1}}, \qquad k_n = \mu^- + \frac{\omega}{4} + \frac{\omega}{2^{n+2}} ; \qquad n = 0, 1, 2, \ldots$$

and construct the family of nested and shrinking cylinders

$$Q_n = Q(d^* R_n^p, c_0 R_n).$$

Write the energy estimate (7.9) for the functions $(\theta - k_n)_-$, over Q_n, and for smooth cutoff functions $0 \leq \xi_n \leq 1$, defined in Q_n, and such that

$$\begin{cases} \xi_n \equiv 1 \ \text{in} \ Q_{n+1} \ ; \xi_n \equiv 0 \ \text{on} \ \partial_p Q_n \\[2mm] |\nabla \xi_n| \leq \frac{2^{n+2}}{c_0 R} \ ; \qquad 0 < (\xi_n)_t \leq \frac{2^{(n+2)p}}{d^* R^p} . \end{cases}$$

Since $(\theta - k_n)_- \leq \frac{\omega}{2}$, we get

$$\sup_{-d^* R_n^p < t < 0} \int_{K_{c_0 R_n} \times \{t\}} (\theta - k_n)_-^2 \xi_n^p + \iint_{Q_n} |\nabla(\theta - k_n)_- \xi_n|^p$$

$$\leq C\left(\frac{\omega}{2}\right)^p \frac{2^{(n+2)p}}{c_0^p R^p} \iint_{Q_n} \chi_{[\theta < k_n]} + C\left(\frac{\omega}{2}\right) \frac{2^{(n+2)p}}{d^* R^p} \iint_{Q_n} \chi_{[\theta < k_n]}$$

$$\leq C\left(\frac{\omega}{2}\right)^p \frac{2^{(n+2)p}}{R^p} \left\{ c_0^{-p} + (d^*)^{-1} \left(\frac{\omega}{2}\right)^{1-p} \right\} \iint_{Q_n} \chi_{[\theta < k_n]}.$$

Now observe that, since $1 < p < 2$, and $(\theta - k_n)_- \leq \frac{\omega}{2}$,

$$(\theta - k_n)_-^2 = (\theta - k_n)_-^{(1-p)(2-p)}(\theta - k_n)_-^{p(2-p)}(\theta - k_n)_-^p$$

$$\geq \left(\frac{\omega}{2}\right)^{(1-p)(2-p)}(\theta - k_n)_-^{p(2-p)}(\theta - k_n)_-^p = d^*(\theta - k_n)_-^{p(2-p)}(\theta - k_n)_-^p.$$

Introducing the level

$$k_{n+1} < \bar{k}_n := \frac{k_n + k_{n+1}}{2} < k_n,$$

we have

$$\int_{K_{c_0 R_n}} (\theta - k_n)_-^2 \xi_n^p \geq d^* \int_{K_{c_0 R_n}} (\theta - k_n)_-^{p(2-p)}(\theta - k_n)_-^p \xi_n^p$$

$$\geq d^* \left(k_n - \bar{k}_n\right)^{p(2-p)} \int_{K_{c_0 R_n}} (\theta - \bar{k}_n)_-^p \xi_n^p$$

$$= d^* \left(\frac{\omega}{2}\right)^{p(2-p)} \left(\frac{1}{2^{n+3}}\right)^{p(2-p)} \int_{K_{c_0 R_n}} (\theta - \bar{k}_n)_-^p \xi_n^p.$$

Consequently,

$$\sup_{-d^* R_n^p < t < 0} \int_{K_{c_0 R_n} \times \{t\}} (\theta - \bar{k}_n)_-^p \xi_n^p$$

$$+ \frac{1}{d^*} \left(\frac{2}{\omega}\right)^{p(2-p)} 2^{(n+3)p(2-p)} \iint_{Q_n} |\nabla(\theta - \bar{k}_n)_- \xi_n|^p$$

$$\leq C\left(\frac{\omega}{2}\right)^p \frac{2^{(n+2)p}}{d^* R^p} 2^{(n+3)p(2-p)} \left\{ \left(\frac{\omega}{2}\right)^{p(p-2)} c_0^{-p} \right.$$

$$\left. + \left(\frac{\omega}{2}\right)^{p(p-2)+(1-p)}(d^*)^{-1} \right\} \iint_{Q_n} \chi_{[\theta < k_n]}$$

$$= C\left(\frac{\omega}{2}\right)^p \frac{2^{(3-p)pn}}{d^* R^p} 2^{p(2+3(2-p))} \left\{ \left(\frac{2}{B}\right)^{p(2-p)} + \frac{2}{\omega} \right\} \iint_{Q_n} \chi_{[\theta < k_n]}$$

$$\leq \frac{C}{\omega} \left(\frac{\omega}{2}\right)^p \frac{2^{(3-p)pn}}{d^* R^p} 2^{p(2+3(2-p))} \left\{ \left(\frac{2}{B}\right)^{p(2-p)} + 2 \right\} \iint_{Q_n} \chi_{[\theta < k_n]}$$

since $\omega \leq 1$.

Consider the change of variables

$$y = \frac{x}{c_0}, \qquad z = \frac{t}{d^*}$$

and define the new functions

$$\bar{\theta}(y, z) = \theta(x, t), \qquad \bar{\xi}_n(y, z) = \xi_n(x, t).$$

Denoting the new variables again by (x, t) and defining

$$A_n = \int_{-R_n^p}^0 |A_n(t)| \, dt, \qquad A_n(t) = \{x \in K_{R_n} : \bar{\theta}(x, t) < k_n\},$$

the above inequality reads

$$\sup_{-R_n^p < t < 0} \int_{K_{R_n}} (\bar{\theta} - \bar{k}_n)_-^p \bar{\xi}_n^p$$

$$+ \left(\frac{2}{\omega}\right)^{p(2-p)} 2^{(n+3)p(2-p)} \iint_{Q(R_n^p, R_n)} \left|\nabla(\bar{\theta} - \bar{k}_n)_- \bar{\xi}_n\right|^p$$

$$\leq \frac{C}{\omega} \left(\frac{\omega}{2}\right)^p \frac{2^{(3-p)pn}}{R^p} 2^{p(2+3(2-p))} \left\{\left(\frac{2}{B}\right)^{p(2-p)} + 2\right\} A_n.$$

Recalling once again that $\omega \leq 1$, and the properties of ξ_n, we conclude that

$$\left\|(\bar{\theta} - \bar{k}_n)_-\right\|_{V^p(Q(R_{n+1}^p, R_{n+1}))}^p$$

$$\leq \frac{C}{\omega} \left(\frac{\omega}{2}\right)^p \frac{2^{(3-p)pn}}{R^p} 2^{p(2+3(2-p))} \left\{\left(\frac{2}{B}\right)^{p(2-p)} + 2\right\} A_n.$$

Now, on the one hand, we have

$$\iint_{Q(R_{n+1}^p, R_{n+1})} (\bar{\theta} - \bar{k}_n)_-^p \geq (\bar{k}_n - k_{n+1})^p A_{n+1} = \left(\frac{\omega}{2}\right)^p \frac{1}{2^{p(n+3)}} A_{n+1}$$

and, on the other hand, using Theorem 2.11,

$$\iint_{Q(R_{n+1}^p, R_{n+1})} (\bar{\theta} - \bar{k}_n)_-^p$$

$$\leq C \left|[\bar{\theta} < \bar{k}_n] \cap Q(R_{n+1}^p, R_{n+1})\right|^{\frac{p}{d+p}} \left\|(\bar{\theta} - \bar{k}_n)_-\right\|_{V^p(Q(R_{n+1}^p, R_{n+1}))}^p$$

$$\leq C A_n^{\frac{p}{d+p}} \left\|(\bar{\theta} - \bar{k}_n)_-\right\|_{V^p(Q(R_{n+1}^p, R_{n+1}))}^p$$

Combining these two inequalities with the previous one, we get

$$A_{n+1} \leq \frac{C}{\omega} \frac{2^{p(5+3(2-p))}}{R^p} \left\{\left(\frac{2}{B}\right)^{p(2-p)} + 2\right\} 2^{(4-p)pn} A_n^{1+\frac{p}{d+p}}.$$

Defining

$$Y_n = \frac{A_n}{|Q(R_n^p, R_n)|}$$

and noting that

$$\frac{|Q(R_n^p, R_n)|^{1+\frac{p}{d+p}}}{|Q(R_{n+1}^p, R_{n+1})|} \le 2^{d+2p} R^p$$

we arrive at

$$Y_{n+1} \le \frac{C}{\omega} \left\{ \left(\frac{2}{B}\right)^{p(2-p)} + 2 \right\} 2^{(4-p)pn} Y_n^{1+\frac{p}{d+p}}.$$

Using Lemma 2.9, we conclude that if

$$Y_0 \le \left(\frac{C}{\omega} \left\{ \left(\frac{2}{B}\right)^{p(2-p)} + 2 \right\} \right)^{\frac{-(d+p)}{p}} 2^{-p(4-p)\left(\frac{d+p}{p}\right)^2}$$

then $Y_n \to 0$ as $n \to \infty$. Since

$$\left\{ \left(\frac{2}{B}\right)^{p(2-p)} + 2 \right\}^{\frac{d+p}{p}} \le 2^{\frac{d}{p}} \left\{ \left(\frac{2}{B}\right)^{(2-p)(d+p)} + 2^{\frac{d+p}{p}} \right\}$$

$$\le 2^{\frac{d}{p}} \left\{ 1 + 2^{\frac{d+p}{p}} \right\},$$

if we take

$$Y_0 \le C^{-\frac{d+p}{p}} 2^{-p(4-p)\left(\frac{d+p}{p}\right)^2} \omega^{\frac{d+p}{p}} 2^{-\frac{d}{p}} \left\{ 1 + 2^{\frac{d+p}{p}} \right\}^{-1} = C\omega^{\frac{d+p}{p}} =: \nu_0 \quad (7.22)$$

we obtain $Y_n \to 0$ as $n \to \infty$, which implies that $A_n \to 0$ as $n \to \infty$. But

$$Y_0 = \frac{\int_{-R^p}^{0} \left| x \in K_R : \bar{\theta}(x,t) < \mu^- + \frac{\omega}{2} \right| dt}{|Q(R^p, R)|},$$

so (7.22) is our hypothesis (7.20). Since $R_n \searrow \frac{R}{2}$ and $k_n \searrow \mu^- + \frac{\omega}{4}$, having $A_n \to 0$ as $n \to \infty$, means that

$$\left| (x,t) \in Q\left(\left(\frac{R}{2}\right)^p, \frac{R}{2} \right) : \bar{\theta}(x,t) \le \mu^- + \frac{\omega}{4} \right| = 0$$

that is, going back to the original variables and function,

$$\theta(x,t) > \mu^- + \frac{\omega}{4}, \quad \text{a.e. } (x,t) \in Q\left(d^* \left(\frac{R}{2}\right)^p, c_0 \frac{R}{2} \right).$$

□

Consider now the time level $-t_* = \bar{t} - d^*(\frac{R}{2})^p$. From the conclusion of Lemma 7.9, we have

$$\theta(x, -t_*) > \mu^- + \frac{\omega}{4}, \quad \text{a.e. } x \in K_{c_0 \frac{R}{2}}.$$

We will use this time level as an initial condition to bring the information up to $t = 0$, and therefore to obtain an analogous inequality in a full cylinder of the type $Q(\tau, c_0\rho)$. A first step in this direction is given by the following result.

Lemma 7.10. *Assume that (7.17) holds for some $\bar{t} \in \mathcal{I}(\omega)$ and for all $\bar{x} \in K_{R(\omega)}$. Given $\nu_1 \in (0,1)$, there exists $s_1 \in \mathbb{N}$, depending only on the data and A, such that, if $B \geq 2^{s_1}$, then*

$$\left|x \in K_{c_0 \frac{R}{4}} : \theta(x,t) < \mu^- + \frac{\omega}{2^{s_1}}\right| < \nu_1 \left|K_{c_0 \frac{R}{4}}\right|, \qquad \forall t \in (-t_*, 0).$$

Proof. Consider the cylinder $Q(t_*, c_0\frac{R}{2})$ and write the logarithmic estimate (7.11) over this cylinder, for the function $(\theta - k)_-$, with

$$k = \mu^- + \frac{\omega}{4} \qquad \text{and} \qquad c = \frac{\omega}{2^{n+2}},$$

where n is to be chosen. Defining

$$k - \theta \leq H_{\theta,k}^- := \operatorname*{ess\,sup}_{Q(t_*, c_0\frac{R}{2})} (\theta - k)_- \leq \frac{\omega}{4}$$

If $H_{\theta,k}^- \leq \frac{\omega}{8}$, the result is trivial for the choice $s_1 = 3$. Assuming $H_{\theta,k}^- > \frac{\omega}{8}$, recall from section 2.3 that the logarithmic function $\psi^- = \psi^-(\theta)$, introduced in (2.7), is well defined and satisfies the estimates

$$\psi^- \leq \ln(2^n) = n \ln 2, \qquad \text{because} \qquad \frac{H_{\theta,k}^-}{H_{\theta,k}^- + \theta - k + c} \leq \frac{\frac{\omega}{4}}{\frac{\omega}{2^{n+2}}} = 2^n;$$

$$|(\psi^-)'|^{2-p} = \left(\frac{1}{H_{\theta,k}^- + \theta - k + c}\right)^{2-p}$$

$$= \left(H_{\theta,k}^- + \theta - k + c\right)^{(p-1)(2-p)} \left(\frac{1}{H_{\theta,k}^- + \theta - k + c}\right)^{p(2-p)}$$

$$\leq \left(\frac{\omega}{2}\right)^{(p-1)(2-p)} \left(\frac{1}{c}\right)^{p(2-p)} = (d^*)^{-1}\left(\frac{2^{n+2}}{\omega}\right)^{p(2-p)},$$

since $c \leq H_{\theta,k}^- + \theta - k + c \leq H_{\theta,k}^- < \frac{\omega}{2}$ and $0 < p - 1 < 1$.

Take as a cutoff function $x \to \xi(x)$ (independent of t), defined in $K_{c_0\frac{R}{2}}$, and satisfying

$$\begin{cases} 0 \leq \xi \leq 1 & \text{in } K_{c_0\frac{R}{2}} \\[2mm] \xi \equiv 1 & \text{in } K_{c_0\frac{R}{4}} \\[2mm] |\nabla\xi| \leq \frac{4}{c_0 R}. \end{cases}$$

The logarithmic estimate takes the form

$$\sup_{-t_* < t < 0} \int_{K_{co\frac{R}{2}}} (\psi^-)^2 \xi^p \le C \int_{-t_*}^0 \int_{K_{co\frac{R}{2}}} \psi^- |(\psi^-)'|^{2-p} |\nabla \xi|^p$$

$$\le Cn \ln 2 \, (d^*)^{-1} \left(\frac{2^{n+2}}{\omega} \right)^{p(2-p)} \frac{4^p}{c_0^p R^p} \left| K_{co\frac{R}{2}} \right| t_*,$$

since $\theta(x, -t_*) > k$ in the cube $K_{co\frac{R}{2}}$ which implies that

$$\psi^-(x, -t_*) = 0, \quad \text{for } x \in K_{co\frac{R}{2}}.$$

Recalling that $t_* < a_0 R^p$, and taking $B \ge 2^{n+2}$, we get

$$\sup_{-t_* < t < 0} \int_{K_{co\frac{R}{2}}} (\psi^-)^2 \xi^p \le C A^{(p-1)(2-p)} n \left| K_{co\frac{R}{4}} \right|.$$

We estimate from below the left hand side of the above inequality integrating over the smaller set

$$\left\{ x \in K_{co\frac{R}{4}} : \theta(x, t) < \mu^- + \frac{\omega}{2^{n+2}} \right\} \subset K_{co\frac{R}{2}}.$$

In this set, $\xi \equiv 1$ and

$$\psi^- = \ln \left(\frac{H_{\theta, k}^-}{H_{\theta, k}^- + \theta - k + c} \right) \ge \ln \left(\frac{H_{\theta, k}^- - \frac{\omega}{4} + \frac{\omega}{4}}{H_{\theta, k}^- - \frac{\omega}{4} + \frac{\omega}{2^{n+1}}} \right) \ge (n-1) \ln 2$$

since $H_{\theta, k}^- - \frac{\omega}{4} \le 0$. Then, for all $t \in (-t_*, 0)$,

$$\left| x \in K_{co\frac{R}{4}} : \theta(x, t) < \mu^- + \frac{\omega}{2^{n+2}} \right| \le C A^{(p-1)(2-p)} \frac{n}{(n-1)^2} \left| K_{co\frac{R}{4}} \right|.$$

Choosing $n > 1 + \frac{2C}{\nu_1} A^{(p-1)(2-p)}$ we get $C A^{(p-1)(2-p)} \frac{n}{(n-1)^2} < \nu_1$ and the result is proved for the choice $s_1 = n + 2$.   □

Remark 7.11. Note that s_1 can only depend on ω through A; it will be shown that A can be determined independently of ω. Observe also that the ϵ dependency appearing in the original logarithmic estimate has been overcome.

We are now in position to prove the main result of this section.

Proposition 7.12. *There exists $s_1 \in \mathbb{N}$, depending only on the data and A, such that, if (7.17) holds for some $\bar{t} \in I(\omega)$ and for all $\bar{x} \in K_{R(\omega)}$, then*

$$\theta(x, t) > \mu^- + \frac{\omega}{2^{s_1+1}}, \quad \text{a.e. } (x, t) \in Q\left(t_*, c_0 \frac{R}{8} \right). \tag{7.23}$$

Proof. Consider the decreasing sequence of real numbers

$$R_n = \frac{R}{8} + \frac{R}{2^{n+3}}, \qquad n = 0, 1, \ldots$$

and construct the family of nested and shrinking cylinders $Q_n = Q(t_*, c_0 R_n)$, where t_* is given as before. Write the energy estimates (7.9) for the functions $(\theta - k_n)_-$ over Q_n, with

$$k_n = \mu^- + \frac{\omega}{2^{s_1+1}} + \frac{\omega}{2^{s_1+1+n}},$$

and choosing piecewise smooth cutoff functions $\xi_n(x)$ defined in $K_{c_0 R_n}$ and satisfying, for $n = 0, 1, 2, \ldots$,

$$\begin{cases} 0 \leq \xi_n \leq 1 & \text{in} \quad K_{c_0 R_n} \\ \\ \xi_n \equiv 1 & \text{in} \quad K_{c_0 R_{n+1}} \\ \\ |\nabla \xi_n| \leq \frac{2^{n+4}}{c_0 R}. \end{cases}$$

Since, for all $n = 0, 1, 2, \ldots$,

$$\theta(x, -t_*) > \mu^- + \frac{\omega}{4} > k_n, \qquad \text{for} \quad x \in K_{c_0 \frac{R}{2}} \supset K_{c_0 R_n}$$

we have

$$\int_{K_{c_0 R_n}} (\theta(\cdot, -t_*) - k_n)_- \, \xi_n^p = 0,$$

and consequently the energy estimates read

$$\sup_{-t_* < t < 0} \int_{K_{c_0 R_n}} (\theta - k_n)_-^2 \xi_n^p + \iint_{Q_n} |\nabla(\theta - k_n)_- \xi_n|^p$$

$$\leq C \iint_{Q_n} (\theta - k_n)_-^p |\nabla \xi_n|^p \leq C \left(\frac{\omega}{2^{s_1}}\right)^p \frac{2^{(n+4)p}}{c_0^p R^p} \iint_{Q_n} \chi_{[\theta < k_n]}.$$

Reasoning as in the proof of Lemma 7.9, we estimate from below the left hand side in the following way: letting

$$k_{n+1} < \bar{k}_n := \frac{k_{n+1} + k_n}{2} < k_n$$

then, since $1 < p < 2$, $(\theta - k_n)_- \leq \frac{\omega}{2^{s_1}}$ and $t_* \leq a_0 R^p$, we have

$$(\theta - k_n)_-^2 \geq \left(\frac{\omega}{2^{s_1}}\right)^{(1-p)(2-p)} \frac{(\frac{R}{2})^p}{t_*} \frac{t_*}{(\frac{R}{2})^p} \left(\frac{\omega}{2^{s_1}}\right)^{p(2-p)} 2^{(n+3)p(p-2)} (\theta - \bar{k}_n)_-^p$$

$$\geq 2^{-p} \left(\frac{\omega}{2^{s_1}}\right)^{(1-p)(2-p)} \left(\frac{\omega}{A}\right)^{(p-1)(2-p)} \left(\frac{\omega}{2^{s_1}}\right)^{p(2-p)} 2^{(n+3)p(p-2)} \frac{t_*}{(\frac{R}{2})^p}(\theta - \bar{k}_n)_-^p$$

$$\geq 2^{-p} \left(\frac{2^{s_1}}{A}\right)^{(p-1)(2-p)} \left(\frac{\omega}{B}\right)^{p(2-p)} 2^{(n+3)p(p-2)} \frac{t_*}{(\frac{R}{2})^p}(\theta - \bar{k}_n)_-^p$$

$$\geq c_0^{-p} 2^{(n+3)p(p-2)} \frac{t_*}{(\frac{R}{2})^p}(\theta - \bar{k}_n)_-^p$$

if we choose $B \geq 2^{s_1}$ and s_1 such that

$$s_1 > \log_2 A + \frac{p}{(p-1)(2-p)}.$$

Therefore the above integral inequality takes the form

$$\sup_{-t_*<t<0} \int_{K_{c_0 R_n}} (\theta - \bar{k}_n)_-^p \xi_n^p + c_0^p 2^{(n+3)p(2-p)} \frac{(\frac{R}{2})^p}{t_*} \int\int_{Q_n} |\nabla(\theta - \bar{k}_n)_- \xi_n|^p$$

$$\leq C \left(\frac{\omega}{2^{s_1}}\right)^p \frac{2^{(3-p)pn}}{(\frac{R}{2})^p} 2^{3p(3-p)} \frac{(\frac{R}{2})^p}{t_*} \int\int_{Q_n} \chi_{[\theta<k_n]}.$$

Introducing the change of variables

$$y = \frac{x}{c_0}, \qquad z = \left(\frac{R}{2}\right)^p \frac{t}{t_*}$$

and defining the new functions

$$\bar{\theta}(y, z) = \theta(x, t), \qquad \bar{\xi}_n(y) = \xi_n(x),$$

we write the inequality in the new variables (again denoted by (x,t)), obtaining

$$\sup_{-(\frac{R}{2})^p<t<0} \int_{K_{R_n}} (\bar{\theta} - \bar{k}_n)_-^p \bar{\xi}_n^p + c_0^p 2^{(n+3)p(2-p)} \int\int_{Q((\frac{R}{2})^p, R_n)} |\nabla(\bar{\theta} - \bar{k}_n)_- \bar{\xi}_n|^p$$

$$\leq C \left(\frac{\omega}{2^{s_1}}\right)^p \frac{2^{(3-p)pn}}{(\frac{R}{2})^p} 2^{3p(3-p)} \int\int_{Q((\frac{R}{2})^p, R_n)} \chi_{[\bar{\theta}<k_n]}.$$

Defining

$$A_n = \int_{-(\frac{R}{2})^p}^0 |A_n(t)|\, dt, \qquad A_n(t) = \{x \in K_{R_n} : \bar{\theta}(x, t) < k_n\}$$

and recalling the definition of c_0 and that $\omega \leq 1$, we arrive at

$$\|(\bar{\theta} - \bar{k}_n)_-\|_{V^p(Q((\frac{R}{2})^p, R_{n+1}))}^p \leq C \left(\frac{\omega}{2^{s_1}}\right)^p \frac{2^{(3-p)pn}}{(\frac{R}{2})^p} 2^{3p(3-p)} A_n.$$

Now, since

$$\left(\frac{\omega}{2^{s_1}}\right)^p \frac{1}{2^{(n+3)p}} A_{n+1} = (\bar{k}_n - k_{n+1})^p A_{n+1}$$

$$\leq \iint_{Q(((\frac{R}{2})^p, R_{n+1})} (\bar{\theta} - \bar{k}_n)^p_- \leq C A_n^{\frac{p}{d+p}} \left\|(\bar{\theta} - \bar{k}_n)_-\right\|^p_{V^p(Q((\frac{R}{2})^p, R_{n+1}))},$$

we get

$$A_{n+1} \leq C \frac{2^{(4-p)pn}}{(\frac{R}{2})^p} 2^{3p(4-p)} A_n^{1+\frac{p}{d+p}}.$$

Define $Y_n = \frac{A_n}{|Q((\frac{R}{2})^p, R_n)|}$. Due to the fact that

$$\frac{|Q((\frac{R}{2})^p, R_n)|^{1+\frac{p}{d+p}}}{|Q((\frac{R}{2})^p, R_{n+1})|} \leq 2^{d+p} \left(\frac{R}{2}\right)^p$$

we get the algebraic inequality

$$Y_{n+1} \leq C 2^{(4-p)pn} Y_n^{1+\frac{p}{d+p}}$$

so, as before, if

$$Y_0 \leq C^{-\frac{d+p}{p}} 2^{-p(4-p)(\frac{d+p}{p})^2} =: \nu_1$$

then $Y_n \to 0$ as $n \to \infty$. Using Lemma 7.10 for this value of ν_1 we conclude that there exists $s_1 \in \mathbb{N}$ such that

$$\left|x \in K_{c_0 \frac{R}{4}} \ : \ \theta < \mu^- + \frac{\omega}{2^{s_1}}\right| < \nu_1 \left|K_{c_0 \frac{R}{4}}\right|, \qquad \forall t \in (-t_*, 0).$$

To conclude, note that

$$Y_0 = \frac{\int_{-(\frac{R}{2})^p}^0 \left|x \in K_{\frac{R}{4}} \ : \ \bar{\theta}(x, t) < \mu^- + \frac{\omega}{2^{s_1}}\right| \, dt}{|Q((\frac{R}{2})^p, \frac{R}{4})|}$$

$$= \frac{(\frac{R}{2})^p}{t_*} \frac{\int_{-t_*}^0 \left|x \in K_{c_0 \frac{R}{4}} \ : \ \theta(x, t) < \mu^- + \frac{\omega}{2^{s_1}}\right| \, dt}{(\frac{R}{2})^p |K_{c_0 \frac{R}{4}}|} \leq \nu_1$$

so $Y_n, A_n \to 0$ as $n \to \infty$. Since

$$R_n \searrow \frac{R}{8} \qquad \text{and} \qquad k_n \searrow \mu^- + \frac{\omega}{2^{s_1+1}},$$

we obtain

$$\theta(x, t) > \mu^- + \frac{\omega}{2^{s_1+1}} \ , \qquad \text{a.e. } (x, t) \in Q\left(t_*, c_0 \frac{R}{8}\right).$$

$\square$

Corollary 7.13. *Assume that (7.17) holds for some $\bar{t} \in \mathcal{I}(\omega)$ and for all $\bar{x} \in K_{\mathcal{R}(\omega)}$. There exists a constant $\sigma_0 \in (0,1)$, depending only on the data and A, such that*

$$\operatorname*{ess\ osc}_{Q(d^*(\frac{R}{2})^p, c_0 \frac{R}{8})} \theta \le \sigma_0 \omega. \tag{7.24}$$

Proof. The proof is trivial, recalling that

$$-t_* = \bar{t} - d^* \left(\frac{R}{2}\right)^p < -d^* \left(\frac{R}{2}\right)^p$$

from which follows $Q\left(t_*, c_0 \frac{R}{8}\right) \supset Q\left(d^*(\frac{R}{2})^p, c_0 \frac{R}{8}\right)$. We obtain

$$\sigma_0 = 1 - \frac{1}{2^{s_1+1}}.$$

$\square$

7.5 The Effect of the Singularity in the Principal Part

If the first alternative does not hold then the second alternative is in force. We will show that, in this case, we can achieve a conclusion similar to (7.24). Note that the constant ν_0 has already been determined and is given by (7.22).

Lemma 7.14. *Fix $\bar{t} \in \mathcal{I}(\omega)$ and $\bar{x} \in K_{\mathcal{R}(\omega)}$ for which (7.18) holds. There exists a time level*

$$t_0 \in \left[\bar{t} - d^* R^p, \bar{t} - \frac{\nu_0}{2} d^* R^p\right] \tag{7.25}$$

such that

$$\left| x \in [\bar{x} + K_{d_* R}] \ : \ \theta(x, t_0) > \mu^+ - \frac{\omega}{2} \right| \le \left(\frac{1 - \nu_0}{1 - \frac{\nu_0}{2}}\right) |K_{d_* R}|. \tag{7.26}$$

Proof. Suppose not. Then, for all $t \in [\bar{t} - d^* R^p, \bar{t} - \frac{\nu_0}{2} d^* R^p]$,

$$\left| (x, t) \in [(\bar{x}, \bar{t}) + Q(d^* R^p, d_* R)] \ : \ \theta(x, t) > \mu^+ - \frac{\omega}{2} \right|$$

$$\ge \int_{\bar{t} - d^* R^p}^{\bar{t} - \frac{\nu_0}{2} d^* R^p} \left| x \in [\bar{x} + K_{d_* R}] \ : \ \theta(x, \tau) > \mu^+ - \frac{\omega}{2} \right| d\tau$$

$$> \left(\frac{1 - \nu_0}{1 - \frac{\nu_0}{2}}\right) |K_{d_* R}| \left(1 - \frac{\nu_0}{2}\right) d^* R^p = (1 - \nu_0) |Q(d^* R^p, d_* R)|$$

which contradicts (7.18).

$\square$

This result tells us that, at the time level t_0, the portion of the cube $[\bar{x} + K_{d_* R}]$ where $\theta(x)$ is near its supremum is small. In what follows we prove that the same happens at all time levels near the top of the cylinder $[(\bar{x}, \bar{t}) + Q(d^* R^p, d_* R)]$.

Lemma 7.15. *There exists $s_2 \in \mathbb{N}$, depending only on the data and ω, such that*

$$\left| x \in [\bar{x} + K_{d_* R}] : \theta(x, t) > \mu^+ - \frac{\omega}{2^{s_2}} \right| < \left(1 - \left(\frac{\nu_0}{2} \right)^2 \right) |K_{d_* R}|, \qquad (7.27)$$

for all $t \in (t_0, \bar{t})$.

Proof. Assume, without loss of generality, that $\bar{x} = 0$. Consider the logarithmic estimate (7.12) written over the cylinder $K_{d_* R} \times (t_0, \bar{t})$, for the function $(\theta - k)_+$, with

$$k = \mu^+ - \frac{\omega}{2} \qquad \text{and} \qquad c = \frac{\omega}{2^{n_*+1}},$$

where n_* is to be chosen. Assuming that the number n_* has been chosen, we determine the length of the cube $K_{d_* R}$ by choosing

$$d_* = \left(\frac{\omega}{2^{n_*+1}} \right)^{p-2}.$$

In the definition of ψ^+ take

$$\theta - k \leq H_{\theta,k}^+ := \operatorname*{ess\,sup}_{K_{d_* R} \times (t_0, \bar{t})} (\theta - k)_+ \leq \frac{\omega}{2}.$$

If $H_{\theta,k}^+ \leq \frac{\omega}{4}$, the result is trivial for the choice $s_2 = 2$. Assuming $H_{\theta,k}^+ > \frac{\omega}{4}$, and since $H_{\theta,k}^+ - \theta + k + c > 0$, the logarithmic function ψ^+ is well defined and satisfies the estimates

$$\psi^+ \leq n_* \ln 2, \quad \text{since} \quad \frac{H_{\theta,k}^+}{H_{\theta,k}^+ - \theta + k + c} \leq \frac{\frac{\omega}{2}}{\frac{\omega}{2^{n_*+1}}} = 2^{n_*};$$

$$\left[(\psi^+)' \right]^{2-p} = \left(H_{\theta,k}^+ - \theta + k + c \right)^{(p-1)(2-p)} \left(\frac{1}{H_{\theta,k}^+ - \theta + k + c} \right)^{p(2-p)}$$

$$\leq (d^*)^{-1} \left(\frac{2^{n_*+1}}{\omega} \right)^{p(2-p)} = (d^*)^{-1} d_*^p,$$

for the non trivial case $\theta > k + c$.

Take as a cutoff function $x \to \xi(x)$, defined in $K_{d_* R}$, and satisfying

$$\begin{cases} 0 \leq \xi \leq 1 \text{ in } K_{d_* R} \\[2mm] \xi \equiv 1 \text{ in } K_{(1-\sigma)d_* R}, \text{ for some } \sigma \in (0, 1) \\[2mm] |\nabla \xi| \leq (\sigma d_* R)^{-1}. \end{cases}$$

The logarithmic estimates read

$$\sup_{t_0 < t < \bar{t}} \int_{K_{d_*R} \times \{t\}} \left(\psi^+\right)^2 \xi^p \leq \int_{K_{d_*R} \times \{t_0\}} \left(\psi^+\right)^2 \xi^p$$

$$+ C \int_{t_0}^{\bar{t}} \int_{K_{d_*R}} \psi^+ \left[\left(\psi^+\right)'\right]^{2-p} |\nabla \xi|^p$$

and therefore

$$\sup_{t_0 < t < \bar{t}} \int_{K_{d_*R} \times \{t\}} \left(\psi^+\right)^2 \xi^p \leq n_*^2 (\ln 2)^2 \left|x \in K_{d_*R} \ : \ \theta(x, t_0) > k + c\right|$$

$$+ \frac{C}{\sigma^p d_*^p R^p} \, n_* \ln 2 \, (d^*)^{-1} d_*^p |K_{d_*R}| \, (\bar{t} - t_0)$$

$$\leq \left\{ n_*^2 (\ln 2)^2 \left(\frac{1 - \nu_0}{1 - \frac{\nu_0}{2}}\right) + C \frac{n_*}{\sigma^p} \right\} |K_{d_*R}|,$$

using Lemma 7.14, the estimates for ψ presented above, and the fact that $\bar{t} - t_0 \leq d^* R^p$. In order to bound the left hand side from below, we integrate over the smaller set

$$S = \left\{ x \in K_{(1-\sigma)d_*R} \ : \ \theta(x, t) > \mu^+ - \frac{\omega}{2^{n_*+1}} \right\} \subset K_{(1-\sigma)d_*R}$$

where $\xi \equiv 1$ and $\psi^+ \geq (n_* - 1) \ln 2$, since

$$\frac{H_{\theta,k}^+}{H_{\theta,k}^+ - \frac{\omega}{2} + \frac{\omega}{2^{n_*}}} = \frac{H_{\theta,k}^+ - \frac{\omega}{2} + \frac{\omega}{2}}{H_{\theta,k}^+ - \frac{\omega}{2} + \frac{\omega}{2^{n_*}}} \geq 2^{n_*-1},$$

because $H_{\theta,k}^+ - \frac{\omega}{2} \leq 0$. We obtain

$$|S| \leq \left\{ \left(\frac{n_*}{n_* - 1}\right)^2 \left(\frac{1 - \nu_0}{1 - \frac{\nu_0}{2}}\right) + \frac{C}{n_* \sigma^p} \right\} |K_{d_*R}|.$$

Consequently, for all $t \in (t_0, \bar{t})$, we have

$$\left| x \in K_{d_*R} : \theta(x, t) > \mu^+ - \frac{\omega}{2^{n_*+1}} \right| \leq |S| + \left| K_{d_*R} \setminus K_{(1-\sigma)d_*R} \right|$$

$$\leq |S| + d\sigma |K_{d_*R}|$$

$$\leq \left\{ \left(\frac{n_*}{n_* - 1}\right)^2 \left(\frac{1 - \nu_0}{1 - \frac{\nu_0}{2}}\right) + \frac{C}{n_* \sigma^p} + d\sigma \right\} |K_{d_*R}|.$$

Choose σ so small that $d\sigma \leq \frac{3}{8} \nu_0^2$ and then n_* so large that

$$\frac{C}{n_* \sigma^p} \leq \frac{3}{8} \nu_0^2 \qquad \text{and} \qquad \left(\frac{n_*}{n_* - 1}\right)^2 \leq \left(1 - \frac{\nu_0}{2}\right)(1 + \nu_0) =: \beta.$$

With these choices we obtain

$$\left| x \in K_{d_*R} \ : \ \theta(x,t) > \mu^+ - \frac{\omega}{2^{n_*+1}} \right| \leq \left(1 - \left(\frac{\nu_0}{2} \right)^2 \right) |K_{d_*R}|, \quad \forall t \in (t_0, \bar{t})$$

and the result follows with $s_2 = n_* + 1$. □

Remark 7.16. From the choice $\sigma \leq \frac{3}{8d} \nu_0^2$, we see that, in order to satisfy the first condition, it suffices to choose the number n_* such that

$$n_* \geq C \nu_0^{-2(p+1)},$$

where C is a constant depending only on the data. From the second condition on n_* we get

$$n_* \geq \frac{\beta + \sqrt{\beta}}{\beta - 1},$$

and, since $\beta > 1$ (and assuming, without loss of generality, that $\nu_0 \in (0, \frac{1}{2})$),

$$\frac{\beta + \sqrt{\beta}}{\beta - 1} \leq \frac{2\beta}{\beta - 1} = \frac{4}{\nu_0(1 - \nu_0)} + 2 \leq \frac{4}{\nu_0^2} + 2$$

and it suffices to choose

$$n_* \geq \frac{4}{\nu_0^2} + 2.$$

So n_* is to be chosen such that

$$n_* \geq \max \left\{ C \nu_0^{-2(p+1)}, \frac{4}{\nu_0^2} + 2 \right\}.$$

Recalling that $\nu_0 = C \omega^{\frac{d+p}{p}}$, we choose

$$n_* \geq C \omega^{-\alpha}, \qquad \alpha = \frac{2(p+1)(d+p)}{p}$$

and n_* depends on the data and ω.

Remark 7.17. This result determines the value s_2 and consequently d_*, which defines the size of the subcubes $[\bar{x} + K_{d_*R}]$ making up the partition of the full cube $K_{c_0 R}$. It thus have a double scope: to determine a level and a cylinder such that the measure of the set where θ is above such a level can be made small on that particular cylinder.

Since for the different values of $\bar{t}$ we may get cylinders with different axes, we now need to expand to a complete cylinder in space and then use the fact that $\bar{t}$ is an arbitrary element of $\mathcal{I}(\omega)$ to get a reduction of the oscillation in a smaller cylinder centered at the origin. In order to do so, and since the location of $\bar{x}$ within the cube $K_{\mathcal{R}(\omega)}$ is only known qualitatively, we will assume that $\bar{x}$

is the centre of the larger cube $[\bar{x} + K_{8c_0R}]$ which we assume to be contained in $K_{R^{\frac{1}{2}}}$. Indeed, if that is not the case, there is nothing to prove since

$$\omega \leq C(B,p) \, R^{\frac{1}{2(2-p)}}.$$

We then work within the cylinder

$$[\bar{x} + K_{8c_0R}] \times (t_0, \bar{t})$$

which is mapped into $Q_4 = K_4 \times (-4^p, 0)$ through the change of variables

$$y = \frac{x - \bar{x}}{2c_0 R}, \qquad z = 4^p \left(\frac{t - \bar{t}}{\bar{t} - t_0} \right).$$

With this same mapping, the cube $[\bar{x} + K_{d_*R}]$ is mapped into K_{h_0}, where $h_0 = \frac{1}{2} \left(\frac{2^{*2}}{B} \right)^{2-p} < 1$. Define the new function

$$\bar{\theta} = (\theta - \mu^+) \left(\frac{2^{n_*}}{\omega} \right)$$

and observe that $\bar{\theta} \leq 0$ and that, for the new variables and function, (7.27) is now written in the form

$$\left| y \in K_{h_0} \, : \, \bar{\theta}(y,z) < -\frac{1}{2} \right| > \left(\frac{\nu_0}{2} \right)^2 |K_{h_0}|, \quad \forall z \in (-4^p, 0). \tag{7.28}$$

7.5.1 An Equation in Dimensionless Form

The new function satisfies, in the sense of the distributions, an equation similar to that satisfied by θ, namely (denoting again the new variables by (x,t))

$$\partial_t \tilde{\gamma}_\epsilon(\bar{\theta}) - C \operatorname{div} \left(|\nabla \bar{\theta}|^{p-2} \nabla \bar{\theta} \right) = 0 \qquad \text{in} \quad \mathcal{D}'(Q_4) \tag{7.29}$$

where $\tilde{\gamma}_\epsilon$ is such that $\tilde{\gamma}_\epsilon{}'(\bar{\theta}) = \gamma_\epsilon'(\theta)$ and

$$C = \frac{1}{2^{3p}} \omega^{(p-1)(2-p)} \left(\frac{2^{n_*}}{B^p} \right)^{2-p} \frac{(\bar{t} - t_0)}{R^p}.$$

Since $t_0 \in \left[\bar{t} - d^* R^p, \bar{t} - \frac{\nu_0}{2} d^* R^p \right]$,

$$C \in \left[\frac{1}{2^{3p}} \left(\frac{2^{n_*+p-1}}{B^p} \right)^{2-p} \frac{\nu_0}{2}, \frac{1}{2^{3p}} \left(\frac{2^{n_*+p-1}}{B^p} \right)^{2-p} \right].$$

In order to simplify the calculations we will assume, for the time being, that

$$\partial_t \tilde{\gamma}_\epsilon \left(\bar{\theta} \right) \in C \left(-4^p, 0; L^1(K_4) \right) \tag{7.30}$$

and will remove this assumption later. The weak formulation of (7.29) is then
given by

$$\int_{K_4} \partial_t \tilde{\gamma}_\epsilon(\bar{\theta}) \varphi + C \int_{K_4} |\nabla \bar{\theta}|^{p-2} \nabla \bar{\theta} \cdot \nabla \varphi = 0,$$

for all $t \in (-4^p, 0)$ and for all $\varphi \in C(Q_4) \cap C(-4^p, 0; W_0^{1,p}(K_4))$.

Due to the fact that the equation is weakly parabolic and $\bar{\theta}$ is a solution,
$(\bar{\theta} - k)_+$ is a sub-solution and then, for all admissible test functions $\varphi \geq 0$,

$$\int_{K_4} \partial_t \tilde{\gamma}_\epsilon \left[(\bar{\theta} - k)_+ \right] \varphi + C \int_{K_4} |\nabla (\bar{\theta} - k)_+|^{p-2} \nabla (\bar{\theta} - k)_+ \cdot \nabla \varphi \leq 0,$$

for all $t \in (-4^p, 0)$. In this inequality we take

$$\varphi = \frac{\xi^p}{\left(-k(1 - \delta) - (\bar{\theta} - k)_+ \right)^{p-1}}$$

where k and $\delta \in (-1, 0)$ are to be chosen and

$$\xi(x, t) = \xi_1(x) \xi_2(t)$$

is a piecewise smooth cutoff function defined in Q_4 satisfying

$$\begin{cases} 0 \leq \xi \leq 1 \text{ in } Q_4 \,; \quad \xi \equiv 1 \text{ in } Q_2 \,; \quad \xi \equiv 0 \text{ on } \partial_p Q_4 \,; \\ |\nabla \xi_1| \leq 1 \,; \quad 0 \leq (\xi_2)_t \leq 1 \,; \end{cases}$$

and the property that the sets $\{x \in K_4 : \xi_1(x) > -k\}$ are convex for all $k \in (-1, 0)$. Set

$$\psi_k(\bar{\theta}) = \ln \left(\frac{-k(1 - \delta)}{-k(1 - \delta) - (\bar{\theta} - k)_+} \right) \geq 0 \tag{7.31}$$

and

$$\phi_k(\bar{\theta}) = \int_0^{(\bar{\theta} - k)_+} \frac{1}{(-k(1 - \delta) - s)^{p-1}} \, ds \tag{7.32}$$

and observe that

1. for $k, \delta \in (-1, 0)$, the functions φ, ψ_k and ϕ_k are well defined since

$$-k(1 - \delta) - (\bar{\theta} - k)_+ = \begin{cases} -k + k\delta \text{ if } \bar{\theta} \leq k \\ k\delta - \bar{\theta} \text{ if } \bar{\theta} > k \end{cases} \geq 0 \,;$$

2. the graph γ has a jump at $\theta = 0$ so the corresponding graph $\tilde{\gamma}$ has a
jump at

$$\bar{\theta} = -\mu^+ \left(\frac{2^{n_*}}{\cdot \omega} \right) < 0.$$

If we choose n_* such that $-\mu^+\left(\frac{2^{n_*}}{\omega}\right) < -1$, i.e.,

$$n_* > \log_2\left(\frac{\omega}{\mu^+}\right)$$

then, for $k \in (-1, 0)$, the function $(\bar{\theta} - k)_+$ is above the singularity in time and $\tilde{\gamma}_\epsilon' \equiv 1$.

Therefore, for

$$n_* > \max\left\{C\omega^{-\alpha}, \log_2\left(\frac{\omega}{\mu^+}\right)\right\}$$

and $k, \delta \in (-1, 0)$, the weak formulation presented above reads

$$\int_{K_4} \partial_t(\bar{\theta} - k)_+ \varphi + C \int_{K_4} \left|\nabla(\bar{\theta} - k)_+\right|^{p-2} \nabla(\bar{\theta} - k)_+ \cdot \nabla\varphi \leq 0, \qquad (7.33)$$

for all $t \in (-4^p, 0)$.

The first term of (7.33) can be estimated from below as follows

$$\int_{K_4} \partial_t(\bar{\theta} - k)_+ \varphi = \int_{K_4} \partial_t\phi_k(\bar{\theta})\xi^p$$

$$= \frac{d}{dt}\int_{K_4} \phi_k(\bar{\theta})\xi^p - p\int_{K_4} \phi_k(\bar{\theta})\xi^{p-1}\xi_t$$

$$\geq \frac{d}{dt}\int_{K_4} \phi_k(\bar{\theta})\xi^p - \frac{C_1}{2-p}$$

using the conditions on ξ and the fact that

$$\phi_k(\bar{\theta}) = \frac{1}{2-p}\left\{(-k(1-\delta))^{2-p} - (-k(1-\delta) - (\bar{\theta} - k)_+)^{2-p}\right\}$$

$$< \frac{1}{2-p}(-k(1-\delta))^{2-p} < \frac{2^{2-p}}{2-p}, \quad k, \delta \in (-1, 0).$$

For the second term we have

$$\int_{K_4} \left|\nabla(\bar{\theta} - k)_+\right|^{p-2} \nabla(\bar{\theta} - k)_+ \cdot \nabla\varphi$$

$$= p\int_{K_4}\left(\frac{\xi}{-k(1-\delta) - (\bar{\theta} - k)_+}\right)^{p-1}\left|\nabla(\bar{\theta} - k)_+\right|^{p-2}\nabla(\bar{\theta} - k)_+ \cdot \nabla\xi$$

$$+(p-1)\int_{K_4}\left(\frac{|\nabla(\bar{\theta} - k)_+|}{-k(1-\delta) - (\bar{\theta} - k)_+}\right)^p \xi^p.$$

Recalling the definition of ψ_k, we have

$$\nabla\psi_k(\bar{\theta}) = \frac{\nabla(\bar{\theta} - k)_+}{-k(1-\delta) - (\bar{\theta} - k)_+}$$

and, using Young's inequality (ρ is to be chosen), we get from the identity above,

$$\int_{K_4} \left|\nabla(\bar{\theta} - k)_+\right|^{p-2} \nabla(\bar{\theta} - k)_+ \cdot \nabla\varphi$$

$$\geq (p-1)\left[1 - \rho^{\frac{-p}{p-1}}\right]\int_{K_4} \left|\nabla\psi_k(\bar{\theta})\right|^p \xi^p - \rho^p \int_{K_4} |\nabla\xi|^p.$$

Taking $\rho = 2^{\frac{p-1}{p}}$, recalling the conditions on ξ and the bounds on C, we finally obtain

$$\frac{d}{dt}\int_{K_4} \phi_k(\bar{\theta})\xi^p + \tilde{C}_0 \int_{K_4} \left|\nabla\psi_k(\bar{\theta})\right|^p \xi^p \leq \frac{\tilde{C}_1}{2-p}, \qquad (7.34)$$

where $\tilde{C}_0 = \tilde{C}_0(p, d, n_*, B, \omega)$ and $\tilde{C}_1 = \tilde{C}_1(p, d, n_*, B)$. Note that, for all $t \in (-4^p, 0)$,

$$\left|[\psi_k(\bar{\theta}) = 0] \cap [\xi = 1]\right| = \left|[\bar{\theta} \leq k] \cap K_2\right| \geq \left|[\bar{\theta} < k] \cap K_{h_0}\right| > \tilde{\nu}_0$$

using (7.28) for $k = -\frac{1}{2}$ and $\tilde{\nu}_0 = \left(\frac{\nu_0}{2}\right)^2 \left(\frac{2^{s_2}}{B}\right)^{d(2-p)} > 0$. We can then apply Theorem 2.12 to conclude that

$$\int_{K_4} \left|\nabla\psi_k(\bar{\theta})\right|^p \xi^p \geq C \int_{K_4} \psi_k^p(\bar{\theta})\xi^p, \qquad C = C(d, p, \tilde{\nu}_0)$$

and consequently the following result.

Lemma 7.18. *There exist constants C_0 and C_1, that can be determined a priori only in terms of, respectively, p, d, ω, n_*, B and the data, and p, d, n_*, B, such that*

$$\frac{d}{dt}\int_{K_4} \phi_k(\bar{\theta})\xi^p + C_0 \int_{K_4} \psi_k^p(\bar{\theta})\xi^p \leq C_1. \qquad (7.35)$$

This integral inequality will be used to prove an auxiliary proposition which is an important tool for what follows. Introduce the quantities

$$Y_n := \sup_{-4^p \leq t \leq 0} \int_{K_4 \cap [\bar{\theta} > -|\delta|^n]} \xi^p, \qquad n = 0, 1, \ldots \qquad (7.36)$$

Proposition 7.19. *The number ν being fixed, we can find numbers δ, σ, depending only on the data, ω and ν, and independent of ϵ, such that, for $n = 0, 1, 2, \ldots$, either*

$$Y_n \leq \nu$$

or

$$Y_{n+1} \leq \max\{\nu, \sigma Y_n\}.$$

Proof. Take $k = -|\delta|^n$ in (7.35), where $\delta \in (-1, 0)$ is to be chosen. From (7.36), it follows that for every $\rho \in (0, 1)$ there exists $t_0 \in (-4^\rho, 0)$ such that

$$Y_{n+1} - \rho \leq \int_{K_4 \cap [\bar{\theta} > -|\delta|^{n+1}]} \xi^p(\cdot, t_0), \qquad n = 0, 1, 2, \ldots \tag{7.37}$$

After fixing $n \in \mathbb{N}$ and $t_0 \in (-4^\rho, 0)$, one of the following two situations holds: either

$$\frac{d}{dt} \left(\int_{K_4} \phi_{-|\delta|^n}(\bar{\theta}) \xi^p \right)(t_0) \geq 0 \tag{7.38}$$

or

$$\frac{d}{dt} \left(\int_{K_4} \phi_{-|\delta|^n}(\bar{\theta}) \xi^p \right)(t_0) < 0. \tag{7.39}$$

In either case we may assume that $Y_n > \nu$ (otherwise the result is trivial).

Assume that (7.38) holds. Then

$$\int_{K_4} \psi^p_{-|\delta|^n}(\bar{\theta}) \xi^p(\cdot, t_0) \leq C,$$

for a positive constant C, independent of ϵ, where

$$\psi^p_{-|\delta|^n}(\bar{\theta}(x, t_0)) = \ln^p \left(\frac{|\delta|^n(1 - \delta)}{-|\delta|^n \delta - \bar{\theta}(x, t_0)} \right) \geq \ln^p \left(\frac{1 - \delta}{-2\delta} \right).$$

Perform an integration over the smaller set

$$\left\{ x \in K_4 \ : \ \bar{\theta}(x, t_0) > -|\delta|^{n+1} \right\}$$

to get

$$\int_{K_4 \cap [\bar{\theta}(\cdot, t_0) > -|\delta|^{n+1}]} \xi^p(\cdot, t_0) \leq C \ln^{-p} \left(\frac{1 - \delta}{-2\delta} \right).$$

Using (7.37) and taking, without loss of generality, $\rho \in (0, \frac{\nu}{2})$, we obtain

$$Y_{n+1} \leq \frac{\nu}{2} + C \ln^{-p} \left(\frac{1 - \delta}{-2\delta} \right).$$

We take $|\delta|$ so small that

$$C \ln^{-p} \left(\frac{1 - \delta}{-2\delta} \right) \leq \frac{\nu}{2},$$

that is

$$|\delta| = -\delta \leq \frac{1}{2 \exp \left(\frac{2C}{\nu} \right)^{\frac{1}{p}} - 1} \in (0, 1)$$

and the proposition is proved assuming that (7.38) holds.

Now assume that (7.39) holds and define

$$t_* := \sup \left\{ t \in (-4^p, t_0) \ : \ \frac{d}{dt} \left(\int_{K_4} \phi_{-|\delta|^n}(\bar{\theta}) \xi^p \right) \geq 0 \right\}.$$

Then

$$\int_{K_4} \phi_{-|\delta|^n}(\bar{\theta}) \xi^p(\cdot, t_0) \leq \int_{K_4} \phi_{-|\delta|^n}(\bar{\theta}) \xi^p(\cdot, t_*)$$

$$= \int_{K_4} \left(\int_0^{|\delta|^n} \chi_{[(\bar{\theta}+|\delta|^n)_+ > s]} \frac{ds}{(|\delta|^n(1-\delta) - s)^{p-1}} \right) \xi^p(\cdot, t_*)$$

$$= \int_{K_4} \left(\int_0^1 \chi_{[(\bar{\theta}+|\delta|^n)_+ > s|\delta|^n]} \frac{|\delta|^{n(2-p)}}{(1-\delta-s)^{p-1}} \, ds \right) \xi^p(\cdot, t_*)$$

$$= \int_0^1 \frac{|\delta|^{n(2-p)}}{(1-\delta-s)^{p-1}} \left(\int_{K_4 \cap [(\bar{\theta}+|\delta|^n)_+ > s|\delta|^n]} \xi^p(\cdot, t_*) \right) ds.$$

We estimate from above the integral in brackets, for $s \in [0, 1]$. On the one hand we have, using the definition of Y_n,

$$\int_{K_4 \cap [(\bar{\theta}+|\delta|^n)_+ > s|\delta|^n]} \xi^p(\cdot, t_*) \leq \int_{K_4 \cap [\bar{\theta} > -|\delta|^n]} \xi^p(\cdot, t_*) \leq Y_n.$$

On the other hand, from the definition of t_*, we first get

$$\int_{K_4} \psi^p_{-|\delta|^n}(\bar{\theta}) \xi^p(\cdot, t_*) \leq C$$

and then, integrating over the smaller set

$$K_4 \cap [(\bar{\theta} + |\delta|^n)_+ > s|\delta|^n],$$

we obtain

$$\int_{K_4 \cap [(\bar{\theta}+|\delta|^n)_+ > s|\delta|^n]} \xi^p(\cdot, t_*) \leq C \ln^{-p} \left(\frac{1-\delta}{1-\delta-s} \right)$$

since, in this set,

$$\psi^p_{-|\delta|^n}(\bar{\theta}) \geq \ln^p \left(\frac{1-\delta}{1-\delta-s} \right).$$

Then, for all $s \in [0, 1]$,

$$\int_{K_4 \cap [(\bar{\theta}+|\delta|^n)_+ > s|\delta|^n]} \xi^p(\cdot, t_*) \leq \min \left\{ Y_n, C \ln^{-p} \left(\frac{1-\delta}{1-\delta-s} \right) \right\}.$$

Let s_* be such that $Y_n = C \ln^{-p}\left(\frac{1-\delta}{1-\delta-s_*}\right)$, i.e.,

$$s_* = \frac{\exp\left(\frac{C}{Y_n}\right)^{\frac{1}{p}} - 1}{\exp\left(\frac{C}{Y_n}\right)^{\frac{1}{p}}}(1-\delta).$$

For $0 \le s < s_*$

$$C \ln^{-p}\left(\frac{1-\delta}{1-\delta-s}\right) > C \ln^{-p}\left(\frac{1-\delta}{1-\delta-s_*}\right) = Y_n$$

and for $s_* \le s \le 1$

$$C \ln^{-p}\left(\frac{1-\delta}{1-\delta-s}\right) \le C \ln^{-p}\left(\frac{1-\delta}{1-\delta-s_*}\right) = Y_n.$$

Then

$$\int_0^1 \frac{|\delta|^{n(2-p)}}{(1-\delta-s)^{p-1}}\left(\int_{K_4 \cap [(\bar\theta+|\delta|^n)_+ > s|\delta|^n]} \xi^p(\cdot, t_*)\right) ds$$

$$\le \int_0^{s_*} \frac{|\delta|^{n(2-p)}}{(1-\delta-s)^{p-1}} Y_n \, ds + \int_{s_*}^1 \frac{|\delta|^{n(2-p)}}{(1-\delta-s)^{p-1}} C \ln^{-p}\left(\frac{1-\delta}{1-\delta-s}\right) ds$$

$$= |\delta|^{n(2-p)} Y_n \left[\int_0^1 \frac{1}{(1-\delta-s)^{p-1}} \, ds \right.$$

$$\left. - \int_{s_*}^1 \frac{1}{(1-\delta-s)^{p-1}}\left\{1 - \frac{C}{Y_n} \ln^{-p}\left(\frac{1-\delta}{1-\delta-s}\right)\right\} ds \right].$$

Our next goal is to obtain an estimate from below, independent of Y_n, for the second integral on the right hand side of this inequality. We start by noting that

$$s_* < \sigma_0(1-\delta), \qquad \sigma_0 := \frac{\exp\left(\frac{C}{\nu}\right)^{\frac{1}{p}} - 1}{\exp\left(\frac{C}{\nu}\right)^{\frac{1}{p}}}$$

since we are assuming that $Y_n > \nu$, and for $s_* \le s \le 1$

$$0 \le 1 - \frac{C}{Y_n} \ln^{-p}\left(\frac{1-\delta}{1-\delta-s}\right).$$

Therefore

$$\int_{s_*}^1 \frac{1}{(1-\delta-s)^{p-1}}\left\{1 - \frac{C}{Y_n} \ln^{-p}\left(\frac{1-\delta}{1-\delta-s}\right)\right\} ds$$

$$\ge \int_{\sigma_0(1-\delta)}^1 \frac{1}{(1-\delta-s)^{p-1}}\left\{1 - \frac{C}{Y_n} \ln^{-p}\left(\frac{1-\delta}{1-\delta-s}\right)\right\} ds$$

$$\ge \int_{\sigma_0(1-\delta)}^1 \frac{1}{(1-\delta-s)^{p-1}}\left\{1 - \frac{C}{\nu} \ln^{-p}\left(\frac{1-\delta}{1-\delta-s}\right)\right\} ds.$$

We obtain

$$\int_{K_4} \phi_{-|\delta|^n}(\bar{\theta})\xi^p(\cdot,t_0) \leq |\delta|^{n(2-p)}Y_n \left[\int_0^1 \frac{1}{(1-\delta-s)^{p-1}} \, ds \right.$$

$$\left. - \int_{\sigma_0(1-\delta)}^1 \frac{1}{(1-\delta-s)^{p-1}} \left\{1 - \frac{C}{\nu}\ln^{-p}\left(\frac{1-\delta}{1-\delta-s}\right)\right\} ds\right]$$

$$= |\delta|^{n(2-p)}Y_n \left[\int_0^{1-|\delta|} \frac{1}{(1-\delta-s)^{p-1}} \, ds - \left\{-\int_{1-|\delta|}^1 \frac{1}{(1-\delta-s)^{p-1}} \, ds\right.\right.$$

$$\left.\left. + \int_{\sigma_0(1-\delta)}^1 \frac{1}{(1-\delta-s)^{p-1}} \left\{1 - \frac{C}{\nu}\ln^{-p}\left(\frac{1-\delta}{1-\delta-s}\right)\right\} ds\right\}\right]$$

$$= |\delta|^{n(2-p)}Y_n[1 - f(\delta)] \int_0^{1-|\delta|} \frac{1}{(1-\delta-s)^{p-1}} \, ds$$

where

$$f(\delta) \int_0^{1-|\delta|} \frac{1}{(1-\delta-s)^{p-1}} \, ds = -\int_{1-|\delta|}^1 \frac{1}{(1-\delta-s)^{p-1}} \, ds$$

$$+ \int_{\sigma_0(1-\delta)}^1 \frac{1}{(1-\delta-s)^{p-1}} \left\{1 - \frac{C}{\nu}\ln^{-p}\left(\frac{1-\delta}{1-\delta-s}\right)\right\} ds.$$

To get a lower bound on $f(\delta)$ note that

(i) for $\sigma_0(1-\delta) \leq s \leq 1$, $1 - \frac{C}{\nu}\ln^{-p}\left(\frac{1-\delta}{1-\delta-s}\right) \geq 0$;

(ii) $\sigma_0 \leq \sigma_1 := \dfrac{\exp\left(\frac{2C}{\nu}\right)^{\frac{1}{p}} - 1}{\exp\left(\frac{2C}{\nu}\right)^{\frac{1}{p}}}$;

(iii) for $\sigma_1(1-\delta) \leq s \leq 1$, $1 - \frac{C}{\nu}\ln^{-p}\left(\frac{1-\delta}{1-\delta-s}\right) \geq \frac{1}{2}$.

Then

$$f(\delta) \int_0^{1-|\delta|} \frac{1}{(1-\delta-s)^{p-1}} \, ds$$

$$\geq -\int_{1-|\delta|}^1 \frac{1}{(1-\delta-s)^{p-1}} \, ds + \frac{1}{2}\int_{\sigma_1(1-\delta)}^1 \frac{1}{(1-\delta-s)^{p-1}} \, ds$$

$$= \frac{1}{2-p}\left(\frac{1}{2}(-\delta)^{2-p} + \frac{1}{2}(1-\delta)^{2-p}(1-\sigma_1)^{2-p} - (-2\delta)^{2-p}\right)$$

$$\geq \frac{1}{2-p}\left(\frac{1}{2}(1-\delta)^{2-p}(1-\sigma_1)^{2-p} - (-2\delta)^{2-p}\right)$$

and consequently

$$f(\delta) \geq \frac{1}{2}(1 - \sigma_1)^{2-p} - \left(\frac{-2\delta}{1 - \delta}\right)^{2-p}.$$

Choosing $\delta \in (-1, 0)$ such that

$$\left(\frac{-2\delta}{1 - \delta}\right)^{2-p} = \frac{1}{4}(1 - \sigma_1)^{2-p}$$

we get

$$f(\delta) \geq \frac{1}{4}(1 - \sigma_1)^{2-p},$$

and then, for $\sigma := 1 - \frac{1}{4}(1 - \sigma_1)^{2-p} \in (0, 1)$,

$$\int_{K_4} \phi_{-|\delta|^n}(\bar{\theta})\xi^p(\cdot, t_0) \leq \sigma Y_n \int_0^{1-|\delta|} \frac{|\delta|^{n(2-p)}}{(1 - \delta - s)^{p-1}} \, ds.$$

To estimate from below the integral over K_4, we integrate over the smaller set $K_4 \cap [\bar{\theta}(\cdot, t_0) > -|\delta|^{n+1}]$ to get, using (7.37),

$$\int_{K_4} \phi_{-|\delta|^n}(\bar{\theta})\xi^p(\cdot, t_0) \geq \int_{K_4 \cap [\bar{\theta}(\cdot, t_0) > -|\delta|^{n+1}]} \phi_{-|\delta|^n}(\bar{\theta})\xi^p(\cdot, t_0)$$

$$= \int_{K_4 \cap [\bar{\theta}(\cdot, t_0) > -|\delta|^{n+1}]} \left(\int_0^{|\delta|^n + \bar{\theta}} \frac{1}{(|\delta|^n(1 - \delta) - s)^{p-1}} \, ds \right) \xi^p(\cdot, t_0)$$

$$\geq \left(\int_{K_4 \cap [\bar{\theta}(\cdot, t_0) > -|\delta|^{n+1}]} \xi^p(\cdot, t_0) \right) \left(\int_0^{1-|\delta|} \frac{|\delta|^{n(2-p)}}{(1 - \delta - s)^{p-1}} \, ds \right)$$

$$\geq (Y_{n+1} - \rho) \int_0^{1-|\delta|} \frac{|\delta|^{n(2-p)}}{(1 - \delta - s)^{p-1}} \, ds.$$

This and the previous estimate yield, since ρ is arbitrary in $(0, \frac{\nu}{2})$,

$$Y_{n+1} \leq \sigma Y_n,$$

which completes the proof also for the case that (7.39) holds.

We now remove assumption (7.30). If (7.30) does not hold we have to make use of the discrete time derivative in order to obtain the weak formulation of (7.29). This means that, for all $t \in [-4^p + h, 0]$, and $h > 0$ we have

$$\int_{t-h}^t \int_{K_4} \partial_t(\bar{\theta})\varphi + C \int_{t-h}^t \int_{K_4} |\nabla\bar{\theta}|^{p-2}\nabla\bar{\theta} \cdot \nabla\varphi = 0,$$

for all admissible testing functions φ, where C is a positive constant independent of ϵ. Being $(\bar{\theta} - k)_+$, with $k \in (-1, 0)$, a sub-solution of (7.29) and taking φ as before, we obtain, for all $t \in [-4^p + h, 0]$ and $h > 0$,

$$C_2 h \geq \int_{K_4 \times \{t\}} \phi_k(\bar{\theta}) \xi^p - \int_{K_4 \times \{t-h\}} \phi_k(\bar{\theta}) \xi^p + C_1 \int_{t-h}^t \int_{K_4} \psi_k^p(\bar{\theta}) \xi^p.$$

Dividing by h and letting $h \to 0$ we get an integral inequality similar to (7.35):

$$\left(\frac{d}{d\tau}\right)^- \int_{K_4} \phi_k(\bar{\theta}) \xi^p + C_1 \int_{K_4} \psi_k^p(\bar{\theta}) \xi^p \leq C_2$$

where

$$\left(\frac{d}{d\tau}\right)^- \int_{K_4} \phi_k(\bar{\theta}) \xi^p := \limsup_{h \to 0} \frac{1}{h} \left[\int_{K_4 \times \{t\}} \phi_k(\bar{\theta}) \xi^p - \int_{K_4 \times \{t-h\}} \phi_k(\bar{\theta}) \xi^p \right].$$

Define the set

$$S := \left\{ t \in (-4^p, 0) : \left(\frac{d}{d\tau}\right)^- \int_{K_4} \phi_k(\bar{\theta}) \xi^p \geq 0 \right\}$$

and let t_0 be given as in (7.37). If $t_0 \in S$, we proceed as in (7.38). If $t_0 \notin S$, take

$$\bar{t} = \sup\{t \in (-4^p, t_0) : t \in S\} \leq t_0.$$

If $\bar{t} = t_0$, consider a sequence $(t_n)_n$, $t_n \in S$, such that $t_n \to t_0$. Since $t_n \in S$ we get

$$\int_{K_4 \times \{t_n\}} \psi_k^p(\bar{\theta}) \xi^p \leq C.$$

Then

$$\int_{K_4 \times \{t_0\}} \psi_k^p(\bar{\theta}) \xi^p \leq C$$

and we proceed as in (7.38). If $\bar{t} < t_0$ we have

$$\begin{cases} \int_{K_4 \times \{\bar{t}\}} \psi_k^p(\bar{\theta}) \xi^p \leq C \\ \int_{K_4} \phi_k(\bar{\theta}) \xi^p(x, t_0) \leq \int_{K_4} \phi_k(\bar{\theta}) \xi^p(x, \bar{t}) \end{cases}$$

and we reason as in (7.39).   □

7.5.2 Expansion in Space

From Proposition 7.19, we get by iteration

$$Y_n \leq \max\{\nu, \sigma^n Y_0\}, \qquad n = 1, 2, \ldots$$

and since

$$Y_0 = \sup_{-4^p \leq t \leq 0} \int_{K_4 \cap [\bar{\theta}(.,t) > -1]} \xi^p(\cdot, t) \leq |K_4|$$

we obtain

$$Y_n \leq \max\{\nu, \sigma^n |K_4|\} = \max\{\nu, \sigma^n 2^d |K_2|\}, \qquad n = 1, 2, \ldots$$

Take $n = n_0 \in \mathbb{N}$ so large that $\sigma^{n_0} 2^d \leq \nu$. Then

$$Y_{n_0} \leq \max\{\nu, \sigma^{n_0} 2^d |K_2|\} \leq \max\{\nu, \nu |K_2|\} = \nu |K_2|.$$

Recalling the definition of Y_n, as well as the choice of ξ, we obtain

$$Y_{n_0} \geq \sup_{-4^p \leq t \leq 0} \int_{K_2 \cap [\bar{\theta}(.,t) > -|\delta|^{n_0}]} \xi^p(\cdot, t) \geq \left| x \in K_2 : \bar{\theta}(x,t) > -|\delta|^{n_0} \right|,$$

for all $t \in [-2^p, 0]$, and therefore

$$\left| x \in K_2 : \bar{\theta}(x,t) > -|\delta|^{n_0} \right| \leq \nu |K_2|, \quad \forall t \in [-2^p, 0].$$

We have just proven the crucial result towards the expansion to a full cylinder in space.

Lemma 7.20. *Given $\nu \in (0,1)$, there exists $\delta^* \in (0,1)$, depending only on the data, ν and ω, such that*

$$\left| x \in K_2 : \bar{\theta}(x,t) > -\delta^* \right| \leq \nu |K_2|, \qquad \forall t \in [-2^p, 0]. \tag{7.40}$$

To prove the main result of this section, we need to make use of another auxiliary result, which proof is a trivial modification to sub-solutions of the proof of Lemma 4.1 of [14, Ch. 4]. Indeed, being *above the singularity* in time, we are dealing only with powers p and 2 in the energy estimates.

Lemma 7.21. *There exists $\tilde{\nu}$, depending on the data, d and p, and independent of ω and ϵ, such that if θ is a sub-solution of (7.6) in*

$$[(\bar{x}, \bar{t}) + Q_R(m_1, m_2)]$$

satisfying

$$\underset{[(\bar{x}, \bar{t}) + Q_R(m_1, m_2)]}{\text{ess osc}} \theta \leq \omega$$

and

$$\left| (x,t) \in [(\bar{x}, \bar{t}) + Q_R(m_1, m_2)] : \theta(x,t) > \mu^+ - \frac{\omega}{2^m} \right| \leq \tilde{\nu} |Q_R(m_1, m_2)|$$

then

$$\theta(x,t) \leq \mu^+ - \frac{\omega}{2^{m+1}}, \qquad \forall (x,y) \in \left[(\bar{x}, \bar{t}) + Q_{\frac{R}{2}}(m_1, m_2) \right],$$

where $m = m_1 + m_2$, $m_1, m_2 \geq 0$ *and*

$$Q_R(m_1, m_2) = K_{d_1 R} \times \left(-2^{m_2(p-2)} R^p, 0\right), \qquad d_1 = \left(\frac{\omega}{2m_1}\right)^{\frac{p-2}{p}}.$$

We are now in a position to prove the main result of this section.

Proposition 7.22. *Assume the second alternative holds. There exists $s_3 > s_2$ such that*

$$\theta(x, t) \leq \mu^+ - \frac{\omega}{2^{s_3}}, \qquad \text{a.e. } (x, t) \in Q\left(\frac{a_0}{2}\left(\frac{R}{4}\right)^p, c_0 R\right). \tag{7.41}$$

Proof. Note that we are done if we prove that (7.41) holds in the cylinder

$$\left[(\bar{x}, 0) + Q\left(\frac{a_0}{2}\left(\frac{R}{4}\right)^p, 2c_0 R\right)\right] \supseteq Q\left(\frac{a_0}{2}\left(\frac{R}{4}\right)^p, c_0 R\right),$$

independently of the location of $\bar{x}$ in $K_{R(\omega)}$. We will prove that there exists $s_4 > 1$ such that

$$\bar{\theta}(x, t) \leq -\frac{1}{2^{s_4}}, \qquad \text{a.e. } (x, t) \in Q_1 = K_1 \times (-1, 0),$$

and then use the fact that $\bar{t}$ is an arbitrary element of $\mathcal{I}(\omega)$ to get (7.41) for the cylinder

$$\left[(\bar{x}, 0) + Q\left(\frac{a_0}{2}\left(\frac{R}{4}\right)^p, 2c_0 R\right)\right]$$

and a proper choice of A.

In Lemma 7.20, take $\nu = \tilde{\nu}$ from Lemma 7.21 and determine the corresponding $\delta^* = \delta^*(\tilde{\nu})$. Then use Lemma 7.21 for $R = 2$, $\mu^+ = 0$, $\omega = 1$, $m_1 = 0$ and m_2 such that $2^{-m_2} = \delta^*(\tilde{\nu})$, over the cylinders

$$\left[(0, \bar{t}) + K_2 \times (-2^{m_2(p-2)}2^p, 0)\right] = [(0, \bar{t}) + Q_2(0, m_2)]$$

as long as they are contained in Q_2, that is, for $\bar{t}$ satisfying

$$-2^p + 2^{m_2(p-2)}2^p \leq \bar{t} \leq 0.$$

Then

$$\left|(x, t) \in [(0, \bar{t}) + Q_2(0, m_2)] : \bar{\theta}(x, t) > -\frac{1}{2^{m_2}}\right| \leq \tilde{\nu} |Q_2(0, m_2)|$$

for each one of the cylinders $[(0, \bar{t}) + Q_2(0, m_2)]$ (since (7.40) holds for all $t \in [-2^p, 0]$). Therefore we conclude that

$$\bar{\theta}(x, t) \leq -\frac{1}{2^{m_2+1}},$$

a.e. $(x,t) \in [(0,\bar{t}) + Q_1(0, m_2)] = K_1 \times (\bar{t} - 2^{m_2(p-2)}, \bar{t})$,

for all $\bar{t} \in [-2^p + 2^{m_2(p-2)}2^p, 0]$. Since $-2^p + 2^{m_2(p-2)}2^p - 2^{m_2(p-2)} < -1$, we get

$$\bar{\theta}(x,t) \leq -\frac{1}{2^{m_2+1}}, \quad \text{a.e. } (x,t) \in Q_1.$$

Returning to the initial variables and function we arrive at

$$\theta(x,t) \leq \mu^+ - \frac{\omega}{2^{m_2+s_2}}, \quad \text{a.e. } (x,t) \in [\bar{x} + K_{2c_0 R}] \times \left(\bar{t} - \frac{\bar{t}-t_0}{4p}, \bar{t}\right).$$

Since $t_0 \in \left[\bar{t} - d^* R^p, \bar{t} - \frac{\nu_0}{2} d^* R^p\right]$, we have

$$\bar{t} - \frac{\bar{t}-t_0}{4p} \in \left[\bar{t} - d^* \left(\frac{R}{4}\right)^p, \bar{t} - \frac{\nu_0}{2} d^* \left(\frac{R}{4}\right)^p\right]$$

and then

$$\theta(x,t) \leq \mu^+ - \frac{\omega}{2^{m_2+s_2}}, \quad \text{a.e. } (x,t) \in K_{c_0 R} \times \left(\bar{t} - \frac{\nu_0}{2} d^* \left(\frac{R}{4}\right)^p, \bar{t}\right)$$

since $[\bar{x} + K_{2c_0 R}] \supset K_{c_0 R}$, independently of the location of $\bar{x} \in K_{R(\omega)}$.

Now we just have to make use of the arbitrary choice of $\bar{t}$ in (7.16) to conclude that

$$\theta(x,t) \leq \mu^+ - \frac{\omega}{2^{m_2+s_2}},$$

$$\text{a.e. } (x,t) \in K_{c_0 R} \times \left(-(a_0 - d^*)R^p - \frac{\nu_0}{2} d^* \left(\frac{R}{4}\right)^p, 0\right).$$

Take A such that

$$\left(\frac{A}{2}\right)^{(p-1)(2-p)} = \frac{a_0}{d^*} \geq 2. \tag{7.42}$$

Then

$$\frac{a_0}{d^*} \geq \frac{2^{2p+1} - \nu_0}{2^{2p+1} - 1} \iff -(a_0 - d^*)R^p - \frac{\nu_0}{2} d^* \left(\frac{R}{4}\right)^p \leq -\frac{a_0}{2} \left(\frac{R}{4}\right)^p$$

and consequently

$$\theta(x,t) \leq \mu^+ - \frac{\omega}{2^{s_3}}, \quad \text{a.e. } (x,t) \in Q\left(\frac{a_0}{2} \left(\frac{R}{4}\right)^p, c_0 R\right)$$

taking $s_3 = m_2 + s_2$.

$\square$

An immediate consequence of Proposition 7.22 is our final result.

Corollary 7.23. *If the second alternative holds, there exists $\sigma_1 \in (0,1)$, depending only on the data and ω, such that*

$$\underset{Q\left(\frac{a_0}{2}\left(\frac{R}{4}\right)^p, c_0 R\right)}{\text{ess osc}} \quad \theta \leq \sigma_1 \omega. \tag{7.43}$$

Remark 7.24. Note that we have only imposed two conditions on A: $A > 2$ and (7.42). So we can take

$$A = 2^{1 + \frac{1}{(p-1)(2-p)}} > 2$$

and conclude that A is independent of ω.

Remark 7.25. As far as B is concerned, we have

$$B > 2^{s_2} \quad \text{and} \quad B \geq 2^{s_1},$$

where, recalling the choice of A, s_1 satisfies

$$s_1 > \log_2 A + \frac{p}{(p-1)(2-p)} = 1 + \frac{p+1}{(p-1)(2-p)}$$

and

$$s_1 > 3 + \frac{2C}{\nu_1} A^{(p-1)(2-p)} = 3 + \frac{C}{\nu_1} 2^{2+(p-1)(2-p)},$$

and

$$s_2 > \max\left\{1 + C\omega^{-\alpha}, 1 + \log_2\left(\frac{\omega}{\mu^+}\right)\right\}.$$

Summarizing:

$$A = 2^{1 + \frac{1}{(p-1)(2-p)}} > 2$$

$$s_1 > \max\left\{3 + \frac{C}{\nu_1} 2^{2+(p-1)(2-p)}, 1 + \frac{p+1}{(p-1)(2-p)}\right\}$$

$$n_* > \max\left\{C\omega^{-\alpha}, \log_2\left(\frac{\omega}{\mu^+}\right)\right\}, \quad \alpha = \frac{2(p+1)(d+p)}{p}$$

$$B > \max\left\{2^{n_*+1}, 2^{s_1}\right\}.$$

References

1. H.W. Alt and E. DiBenedetto, *Nonsteady flow of water and oil through inhomogeneous porous media*, Ann. Sc. Norm. Sup. Pisa (IV) **12** (1985), 335-392.
2. S.N. Antontsev and S. Shmarev, *A model porous medium equation with variable exponent of nonlinearity: existence, uniqueness and localization properties of solutions*, Nonlinear Anal. **60** (2005), 515-545.
3. D.G. Aronson, *The Porous Medium Equation*, in: Nonlinear Diffusion Problems (Montecatini Terme), pp. 1-46, Lecture Notes in Math. **1224**, C.I.M.E. series, Springer, Berlin, 1986.
4. M. Bendahmane, K.H. Karlsen and J.M. Urbano, *On a two-sidedly degenerate chemotaxis model with volume-filling effect*, Math. Models Methods Appl. Sci. **17** (2007), 783-804.
5. E. Bombieri, *Ennio De Giorgi*, Rend. Suppl. Acc. Lincei **8** (1997), 105-114.
6. M. Burger, M. Di Francesco and Y. Dolak-Struss, *The Keller-Segel model for chemotaxis with prevention of overcrowding: linear vs. nonlinear diffusion*, SIAM J. Math. Anal. **38** (2006), 1288-1315.
7. L.A. Caffarelli and L.C. Evans, *Continuity of the temperature in the two phase Stefan problem*, Arch. Ration. Mech. Anal. **81** (1983), 199-220.
8. Y. Chen, S. Levine and M. Rao, *Variable exponent, linear growth functionals in image restoration*, SIAM J. Appl. Math. **66** (2006), 1383-1406.
9. Y.Z. Chen and E. DiBenedetto, *On the local behaviour of solutions of singular parabolic equations*, Arch. Ration. Mech. Anal. **103** (1988), 319-345.
10. E. DeGiorgi, *Sulla differenziabilitá e l'analiticitá delle estremali degli integrali multipli regolari*, Mem. Accad. Sci. Torino Cl. Sci. Fis. Mat. Natur. **3** (1957), 25-43.
11. E. DiBenedetto, *Continuity of weak solutions to certain singular parabolic equations*, Ann. Mat. Pura Appl. **121** (1982), 131-176.
12. E. DiBenedetto, *On the local behaviour of solutions of degenerate parabolic equations with measurable coefficients*, Ann. Sc. Norm. Super. Pisa **13** (1986), 487-535.
13. E. DiBenedetto, *The flow of two immiscible fluids through a porous medium. Regularity of the saturation*, in: "Theory and Applications of liquid crystals" (Minneapolis, 1985), 123-141, IMA Vol. Math. Appl. **5**, Springer-Verlag, New York, 1987.

14. E. DiBenedetto, *Degenerate Parabolic Equations*, Springer-Verlag, Series Universitext, New York, 1993.

15. E. DiBenedetto, *Parabolic Equations with Multiple Singularities*, in: Proceedings Equadiff 9 (Brno 1997), pp. 25-48, Electronic Publishing House, Stony Brook, 1998.

16. E. DiBenedetto and R. Gariepy, *On the local behavior of solutions for an elliptic-parabolic equation*, Arch. Ration. Mech. Anal. **97** (1987), 1-17.

17. E. DiBenedetto, U. Gianazza and V. Vespri, *Harnack estimates for quasi-linear degenerate parabolic differential equations*, Acta Math., to appear.

18. E. DiBenedetto, U. Gianazza and V. Vespri, *Sub-potential lower bounds for non-negative solutions to certain quasi-linear degenerate parabolic differential equations*, Duke Math. J., to appear.

19. E. DiBenedetto, U. Gianazza and V. Vespri, *Forward, backward and elliptic harnack inequalities for non-negative solutions to certain singular parabolic partial differential equations*, preprint.

20. E. DiBenedetto, U. Gianazza, M. Safonov, J.M. Urbano and V. Vespri, *Harnacks Estimates: Positivity and Local Behavior of Degenerate and Singular Parabolic Equations - Special Issue*, Bound. Value Probl., Hindawi Publ. Corp., 2007.

21. E. DiBenedetto, J.M. Urbano and V. Vespri, *Current issues on singular and degenerate evolution equations*, in: Handbook of Differential Equations, Evolutionary Equations, vol. 1, pp. 169-286, Elsevier, 2004.

22. E. DiBenedetto and V. Vespri, *On the singular equation $\beta(u)_t = \triangle u$*, Arch. Ration. Mech. Anal. **132** (1995), 247-309.

23. U. Gianazza, B. Stroffolini and V. Vespri, *Interior and boundary continuity of the weak solution of the singular equation $(\beta(u))_t = \mathcal{L}u$*, Nonlinear Anal. **56** (2004), 157-183.

24. U. Gianazza and V. Vespri, *Continuity of weak solutions of a singular parabolic equation*, Adv. Differential Equations **8** (2003), 1341-1376.

25. J. Hadamard, *Extension à l'équation de la chaleur d'un théorem de A. Harnack*, Rend. Circ. Mat. Palermo **3** (1954), 337-346.

26. E. Henriques, *Regularity for the porous medium equation with variable exponent: the singular case*, J. Differential Equations, to appear.

27. E. Henriques and J.M. Urbano, *On the doubly singular equation $\gamma(u)_t = \Delta_p u$*, Comm. Partial Differential Equations **30** (2005), 919-955.

28. E. Henriques and J.M. Urbano, *Intrinsic scaling for PDEs with an exponential nonlinearity*, Indiana Univ. Math. J. **55** (2006), 1701-1722.

29. T. Hillen and K. Painter, *Global existence for a parabolic chemotaxis model with prevention of overcrowding*, Adv. in Appl. Math. **26** (2001) 280-301.

30. D. Hoff, *A scheme for computing solutions and interface curves for a doubly-degenerate parabolic equation*, SIAM J. Numer. Anal. **22** (1985), 687-712.

31. A.V. Ivanov, *Uniform Hölder estimates for generalized solutions of quasilinear parabolic equations admitting a doubly degeneracy*, Algebra i Analiz **3** (1991), 139-179.

32. E.F. Keller and L.A. Segel, *Model for chemotaxis*, J. Theor. Biol. **30** (1971), 225-234.

33. S.N. Kruzkov, *On the a priori estimation of solutions of linear parabolic equations and of solutions of boundary value problems for a certain class of quasilinear parabolic equations*, Dokl. Akad. NAUK SSSR **138** (1961), 1005-1008; (English transl.: Soviet Math. Dokl. **2** (1961), 764-767).

34. S.N. Kruzkov, *A priori estimates and certain properties of the solutions of elliptic and parabolic equations of second order*, Math. Sbornik **65** (1968), 522-570; (English. transl.: Amer. Math. Soc. Transl. **2** (68) (1968), 169-220.

35. S.N. Kruzkov and S.M. Sukorjanski, *Boundary value problems for systems of equations of two phase porous flow type: statement of the problems, questions of solvability, justification of approximate methods*, Mat. Sb. **33** (1977), 62-80.

36. T. Kuusi, *Harnack estimates for supersolutions to a nonlinear degenerate parabolic equation*, Helsinki University of Technology, Institute of Mathematics, Research Report A532, 2007.

37. O.A. Ladyzhenskaja, V.A. Solonnikov and N.N. Ural'ceva, *Linear and Quasilinear Equations of Parabolic Type*, Amer. Math. Soc. Transl. Math. Mono., Vol. **23**, Providence, RI, 1968.

38. O.A. Ladyzhenskaja and N.N. Ural'ceva, *Linear and Quasilinear Elliptic Equations*, Academic Press, New York, 1968.

39. J.L. Lions, *Quelques Méthodes de Résolution dés Problèmes aux Limites Nonlineaires*, Dunod, Paris, 1969.

40. A. Meirmanov, *The Stefan Problem*, Walter de Gruyter, Berlin, 1992.

41. C.B. Morrey, *On the solutions of quasi-linear elliptic partial differential equations*, Trans. Amer. Math. Soc. **43** (1938), 126-166.

42. J. Moser, *A new proof of DeGiorgi's theorem concerning the regularity problem for elliptic differential equations*, Comm. Pure Appl. Math. **13** (1960), 457-468.

43. J. Moser, *A Harnack inequality for parabolic differential equations*, Comm. Pure Appl. Math. **17** (1964), 101-134.

44. J. Nash, *Continuity of solutions of parabolic and elliptic equations*, Amer. J. Math. **80** (1958), 931-954.

45. B. Pini, *Sulla soluzione generalizzata di Wiener per il primo problema di valori al contorno nel caso parabolico*, Rend. Sem. Math. Univ. Padova **23** (1954), 422-434.

46. M.M. Porzio and V. Vespri, *Hölder estimates for local solutions of some doubly nonlinear degenerate parabolic equations*, J. Differential Equations **103** (1993), 146-178.

47. M. Růžička, *Electrorheological fluids: modelling and mathematical theory*, Lect. Notes Math. **1748**, Springer-Verlag, Berlin, 2000.

48. P. Sachs, *Continuity of solutions of singular parabolic equations*, Nonlinear Anal. **7** (1983), 387-409.

49. P. Sachs, *The initial and boundary value problem for a class of degenerate parabolic equations*, Comm. Partial Differential Equations **8** (1983), 693-734.

50. J.M. Urbano, *A free boundary problem with convection for the p-Laplacian*, Rend. Mat. Appl. **17** (1997), 1-19.

51. J.M. Urbano, *Continuous solutions for a degenerate free boundary problem*, Ann. Mat. Pura Appl. **178** (2000), 195-224.

52. J.M. Urbano, *A singular-degenerate parabolic problem: regularity up to the Dirichlet boundary*, in: Free Boundary Problems - Theory and Applications I, pp. 399-410, GAKUTO Int. Series Math. Sci. Appl. **13**, 2000.

53. J.M. Urbano, *Hölder continuity of local weak solutions for parabolic equations exhibiting two degeneracies*, Adv. Differential Equations **6** (2001), 327-358.

54. J.M. Urbano, *Regularity for partial differential equations: from De Giorgi-Nash-Moser theory to intrinsic scaling*, in: CIM Bull. **12** (2002), 8-14.

55. J.L. Vázquez, *The Porous Medium Equation: mathematical theory*, Oxford Mathematical Monographs, Oxford University Press, Oxford, 2007.

56. V. Zhikov, *Averaging of functionals of the calculus of variations and elasticity theory*, Izv. Akad. Nauk SSSR Ser. Mat. **50** (1986), 675-710, 877.

57. W.P. Ziemer, *Interior and boundary continuity of weak solutions of degenerate parabolic equations*, Trans. Amer. Math. Soc. **271** (1982), 733-748.

Index

Lecture Notes in Mathematics

For information about earlier volumes
please contact your bookseller or Springer
LNM Online archive: springerlink.com

Vol. 1803: G. Dolzmann, Variational Methods for Crystalline Microstructure – Analysis and Computation (2003)
Vol. 1804: I. Cherednik, Ya. Markov, R. Howe, G. Lusztig, Iwahori-Hecke Algebras and their Representation Theory. Martina Franca, Italy 1999. Editors: V. Baldoni, D. Barbasch (2003)
Vol. 1805: F. Cao, Geometric Curve Evolution and Image Processing (2003)
Vol. 1806: H. Broer, I. Hoveijn. G. Lunther, G. Vegter, Bifurcations in Hamiltonian Systems. Computing Singularities by Gröbner Bases (2003)
Vol. 1807: V. D. Milman, G. Schechtman (Eds.), Geometric Aspects of Functional Analysis. Israel Seminar 2000-2002 (2003)
Vol. 1808: W. Schindler, Measures with Symmetry Properties (2003)
Vol. 1809: O. Steinbach, Stability Estimates for Hybrid Coupled Domain Decomposition Methods (2003)
Vol. 1810: J. Wengenroth, Derived Functors in Functional Analysis (2003)
Vol. 1811: J. Stevens, Deformations of Singularities (2003)
Vol. 1812: L. Ambrosio, K. Deckelnick, G. Dziuk, M. Mimura, V. A. Solonnikov, H. M. Soner, Mathematical Aspects of Evolving Interfaces. Madeira, Funchal, Portugal 2000. Editors: P. Colli, J. F. Rodrigues (2003)
Vol. 1813: L. Ambrosio, L. A. Caffarelli, Y. Brenier, G. Buttazzo, C. Villani, Optimal Transportation and its Applications. Martina Franca, Italy 2001. Editors: L. A. Caffarelli, S. Salsa (2003)
Vol. 1814: P. Bank, F. Baudoin, H. Föllmer, L.C.G. Rogers, M. Soner, N. Touzi, Paris-Princeton Lectures on Mathematical Finance 2002 (2003)
Vol. 1815: A. M. Vershik (Ed.), Asymptotic Combinatorics with Applications to Mathematical Physics. St. Petersburg, Russia 2001 (2003)
Vol. 1816: S. Albeverio, W. Schachermayer, M. Talagrand, Lectures on Probability Theory and Statistics. Ecole d'Eté de Probabilités de Saint-Flour XXX-2000. Editor: P. Bernard (2003)
Vol. 1817: E. Koelink, W. Van Assche (Eds.), Orthogonal Polynomials and Special Functions. Leuven 2002 (2003)
Vol. 1818: M. Bildhauer, Convex Variational Problems with Linear, nearly Linear and/or Anisotropic Growth Conditions (2003)
Vol. 1819: D. Masser, Yu. V. Nesterenko, H. P. Schlickewei, W. M. Schmidt, M. Waldschmidt, Diophantine Approximation. Cetraro, Italy 2000. Editors: F. Amoroso, U. Zannier (2003)
Vol. 1820: F. Hiai, H. Kosaki, Means of Hilbert Space Operators (2003)
Vol. 1821: S. Teufel, Adiabatic Perturbation Theory in Quantum Dynamics (2003)
Vol. 1822: S.-N. Chow, R. Conti, R. Johnson, J. Mallet-Paret, R. Nussbaum, Dynamical Systems. Cetraro, Italy 2000. Editors: J. W. Macki, P. Zecca (2003)
Vol. 1823: A. M. Anile, W. Allegretto, C. Ringhofer, Mathematical Problems in Semiconductor Physics. Cetraro, Italy 1998. Editor: A. M. Anile (2003)
Vol. 1824: J. A. Navarro González, J. B. Sancho de Salas, $\mathscr{C}^\infty$ – Differentiable Spaces (2003)
Vol. 1825: J. H. Bramble, A. Cohen, W. Dahmen, Multiscale Problems and Methods in Numerical Simulations, Martina Franca, Italy 2001. Editor: C. Canuto (2003)
Vol. 1826: K. Dohmen, Improved Bonferroni Inequalities via Abstract Tubes. Inequalities and Identities of Inclusion-Exclusion Type. VIII, 113 p, 2003.

Vol. 1827: K. M. Pilgrim, Combinations of Complex Dynamical Systems. IX, 118 p, 2003.
Vol. 1828: D. J. Green, Gröbner Bases and the Computation of Group Cohomology. XII, 138 p, 2003.
Vol. 1829: E. Altman, B. Gaujal, A. Hordijk, Discrete-Event Control of Stochastic Networks: Multimodularity and Regularity. XIV, 313 p, 2003.
Vol. 1830: M. I. Gil', Operator Functions and Localization of Spectra. XIV, 256 p, 2003.
Vol. 1831: A. Connes, J. Cuntz, E. Guentner, N. Higson, J. E. Kaminker, Noncommutative Geometry, Martina Franca, Italy 2002. Editors: S. Doplicher, L. Longo (2004)
Vol. 1832: J. Azéma, M. Émery, M. Ledoux, M. Yor (Eds.), Séminaire de Probabilités XXXVII (2003)
Vol. 1833: D.-Q. Jiang, M. Qian, M.-P. Qian, Mathematical Theory of Nonequilibrium Steady States. On the Frontier of Probability and Dynamical Systems. IX, 280 p, 2004.
Vol. 1834: Yo. Yomdin, G. Comte, Tame Geometry with Application in Smooth Analysis. VIII, 186 p, 2004.
Vol. 1835: O.T. Izhboldin, B. Kahn, N.A. Karpenko, A. Vishik, Geometric Methods in the Algebraic Theory of Quadratic Forms. Summer School, Lens, 2000. Editor: J.-P. Tignol (2004)
Vol. 1836: C. Năstăsescu, F. Van Oystaeyen, Methods of Graded Rings. XIII, 304 p, 2004.
Vol. 1837: S. Tavaré, O. Zeitouni, Lectures on Probability Theory and Statistics. Ecole d'Eté de Probabilités de Saint-Flour XXXI-2001. Editor: J. Picard (2004)
Vol. 1838: A.J. Ganesh, N.W. O'Connell, D.J. Wischik, Big Queues. XII, 254 p, 2004.
Vol. 1839: R. Gohm, Noncommutative Stationary Processes. VIII, 170 p, 2004.
Vol. 1840: B. Tsirelson, W. Werner, Lectures on Probability Theory and Statistics. Ecole d'Eté de Probabilités de Saint-Flour XXXII-2002. Editor: J. Picard (2004)
Vol. 1841: W. Reichel, Uniqueness Theorems for Variational Problems by the Method of Transformation Groups (2004)
Vol. 1842: T. Johnsen, A. L. Knutsen, K_3 Projective Models in Scrolls (2004)
Vol. 1843: B. Jefferies, Spectral Properties of Noncommuting Operators (2004)
Vol. 1844: K.F. Siburg, The Principle of Least Action in Geometry and Dynamics (2004)
Vol. 1845: Min Ho Lee, Mixed Automorphic Forms, Torus Bundles, and Jacobi Forms (2004)
Vol. 1846: H. Ammari, H. Kang, Reconstruction of Small Inhomogeneities from Boundary Measurements (2004)
Vol. 1847: T.R. Bielecki, T. Björk, M. Jeanblanc, M. Rutkowski, J.A. Scheinkman, W. Xiong, Paris-Princeton Lectures on Mathematical Finance 2003 (2004)
Vol. 1848: M. Abate, J. E. Fornaess, X. Huang, J. P. Rosay, A. Tumanov, Real Methods in Complex and CR Geometry, Martina Franca, Italy 2002. Editors: D. Zaitsev, G. Zampieri (2004)
Vol. 1849: Martin L. Brown, Heegner Modules and Elliptic Curves (2004)
Vol. 1850: V. D. Milman, G. Schechtman (Eds.), Geometric Aspects of Functional Analysis. Israel Seminar 2002-2003 (2004)
Vol. 1851: O. Catoni, Statistical Learning Theory and Stochastic Optimization (2004)
Vol. 1852: A.S. Kechris, B.D. Miller, Topics in Orbit Equivalence (2004)
Vol. 1853: Ch. Favre, M. Jonsson, The Valuative Tree (2004)

Recent Reprints and New Editions